"*The Clean Power Revolution* is o
have ever read in my life. Troy Helming is a true genius. When I
first met Troy Helming, it was like meeting the Thomas Edison of
the 21st Century. Troy brilliantly exposes our energy problems in
the world and then gives real world solutions to the energy crises.
If you are concerned about the quality of your life in the future then
The Clean Power Revolution is a must read. I am so thankful that
Troy took my advice and wrote this soon to be called a classic
contribution to humanity."
Jack Lannom (FL), author of *People First*

"Finally a book has been written that shows the way environmental
and fiscal responsibility can, and do, go hand in hand."
Sky Canyon (CO), former President & Publisher, **New World Library**

"Mr. Helming provides not only an innovative, clean solution to an
age-old problem, but an infinitely practical one as well."
Jim Dallas (NY), Managing Director, **Core Capital Holdings**

"A must read for anyone tired of paying too much at the pump...
this book is based on a solid foundation that builds a plan to get
America free from dependence on foreign oil."
Craig Andres (MO), Professional Journalist & former **TV News Producer**

"The Freedom Plan is mobilizing; even the most conservative
Republicans will take action and switch to clean Green Power."
Jeff Eagle (KS), **US Navy Submarine Force, Ret**

"Full of shocking facts about our energy crisis but gives a hopeful,
provocative solution. This book will motivate both the
environmentalist and business person to action."
Hanh Pham (CA), **MBA student at Presidio World College**

"Troy Helming is a master of communicating environmental benefits in a common-sense, practical style that **everyone** can relate to. Skills like his are critical to aspiring sustainable business leaders, and the *Clean Power Revolution* showcases his revolutionary yet easy-to-understand approach."
Ivan Storck (CA), **Founder** of <u>**SustainableMarketing.com**</u>

"Star Wars III was great, but The Clean Power Revolution should be at the top of your priority list to read this summer. Its time we all stop lying to ourselves about the true economic and environmental havoc that is banging at our doorsteps and become part of the solution. Clearly outlined in a well written manner, The Freedom Plan is a modern day masterpiece that's a gripping and stimulating read. A stunning vision meets meets practical, realistic application that just jumps off the page - Mr. Helming's use of sound facts and years of proven research make the implementation of this plan an undeniably wise course of action and the only real solution to converting our planet to clean power."
Timothy Beckford (FL), **College Student**

"Timely, Easy to read, and filled with Valuable Facts about our current energy policies and a hope filled and bright overview of what Renewable energy can do for us. A call for Action."
Adriana Sanchez Gomez (FL), **Founder/CEO, Energy Consultants**

"Troy's book is a strong wake up call to America. It's power-packed with valuable information. We have a moral duty to take care of the earth and we'd better start making changes right now. People need to ask what they can do today, or there'll be a huge price to pay tomorrow."
Jason G Poulos (FL), **Pastor, Evangelical Church of Miami**

To submit an independent review, send an email to: troy@troyhelming.com
(Please indicate that you provide permission to use your name)

The Clean Power Revolution

By

Troy A. Helming

Printed on recycled and/or eucalyptus paper

**To purchase additional copies of this book,
visit one of these websites:**

www.TroyHelming.com

www.CleanPowerRevolution.org

www.TheFreedomPlan.org

(Or call 913-888-0500 x157)

A large portion of the proceeds earned from the sale of this book is contributed towards construction of clean renewable energy projects in the USA.

Published by Team Reach Inc.

ISBN 0-9767610-0-9

© Team Reach Inc. 2005
4th Printing

DUPLICATION strictly prohibited.

"The Clean Power Revolution"

Contents

Chapter   Title

-     Forward
1   Overview
2   The Automobile
3   Electricity: the Worst Polluter
4   Winds of Change
5   Hydrogen
6   The Freedom Plan
7   Save $20 Trillion over 20 Years
8   "Duh!"
9   Planet Savers
10   Making a Difference
11   An Exciting Future
12   Conclusion
-     About the Author
-     Supporters of The Freedom Plan

To Alysia, my heaven on Earth

Foreword

Written by Craig Andres,
TV News Producer and professional Journalist for two decades.

Wind and Hydrogen can save America $20 Trillion by 2025

You are about to embark on a journey. This book will take you places you never dreamed possible. It's no secret the world oil supply will not last forever. Many are questioning when gas prices will spike again... reaching levels we have never seen. The Clean Power Revolution offers a realistic way to hope. It showcases a Revolution that can change the planet. It describes how you can protect yourself and your family now from rising and volatile energy costs, **while savings $150 per month on more on your own energy bills starting now.** And, it all starts right here.

Read on with confidence. This book is well researched with solid ideas and a plan to both save the planet and get us away from dependence on foreign energy. Furthermore, while reading this book you will ponder the cost of the Middle East military action. From 2003 to 2005 the conflict cost will exceed $300 Billion. And, our men and women are sacrificing themselves to help make the Middle East a stable environment.

With this simple book, you will learn how we can spend less money and avoid sending more troops to the Middle East for direct military conflicts. The Clean Power Revolution and the Freedom Plan will empower you with the tools needed to change our country and even our world.

The Clean Power Revolution says, "...global warming is wreaking havoc in visible, as well as more subtle ways that we may not have discovered yet... The indirect cost to our economy of global warming... could become staggering in less than a decade." But there is hope.

Imagine driving to work in a hydrogen powered vehicle. Imagine putting the Clean Power Revolution to work. Imagine one million Americans taking action that will change the way we create energy forever. It can happen. And you will find it is a very real possibility when you read this book.

Author *Troy Helming* is passionate about energy. *Mr. Helming* believes the hydrogen economy is largely misunderstood by mainstream America. This book will educate the lay person as well as the engineer about the risks and rewards of hydrogen, and why we should embrace this clean gas. In addition to hydrogen, *Mr. Helming* explains how wind energy can and should be harnessed to make the hydrogen economy take wings and soar.

Chapter after chapter, *Mr. Helming* spells out a clear plan of action. You will come away from this book both enlightened and entertained. And you will even be armed with one-liners about clean power you can tell your friends.

The current cost of the 2003 – 2005 military action in the Middle East exceeds $300 Billion and this is just one of the places where costs can be cut when America finally takes full advantage of wind electricity production and starts milling water to produce hydrogen to run the combustion engine. Mr. Helming will demonstrate how America can stop spending money and wasting life to secure oil production, and how America can supply the world with technology and infrastructure to fuel the 800 million internal-combustion engines on the planet with hydrogen.

"There is no doubt that global warming is wreaking havoc in visible, as well as subtle ways that we have not yet discovered, and that this damage is occurring worldwide. Such trends force our nation – and others – to pay an ever increasing percentage of GDP towards the costs of global warming. These costs are both direct, which are easily quantified, and indirect. The indirect costs to our economy of global warming, the true nature of which are only now being discovered and are more difficult to measure, remain elusive. But research suggests that the true cost to our economy of global warming could become staggering in less than a decade." [Excerpt from *The Clean Power Revolution*]

These indirect costs can translate into savings by changing the composition of the energy consumed in America. This starts by replacing coal-burning power plants with wind energy facilities. The amount of money saved on healthcare alone is just one of the many costs that can be avoided by reducing the pollution from coal plants.

"The only way to escape pollution from the burning of fossil fuels is to move to a remote mountain hideaway, grow all your own food locally using organic methods, filter all your water, and stop breathing." [Excerpt]

The hydrogen economy is largely misunderstood by mainstream America. This book will educate the lay person as well as the engineer about the risks and rewards of hydrogen, and why we should embrace this clean gas as our primary energy carrier - today.

"Most of the hydrogen needed to displace all the world's gasoline is *already being produced* for other purposes, including making gasoline. Early adopters of home hydrogen systems, existing and future state and local government incentives for hydrogen systems, and clever marketing companies will combine to accelerate the installation of numerous fuel cell systems in a distributed manner... The businesses that identify these niche opportunities to utilize the performance benefits of hydrogen will quickly differentiate themselves from their competitors in both public relations and profitability." [Excerpt]

America can solve its energy problems completely and eliminate dependence on foreign energy, while making a lot of money all at the same time. Read on... this journey will change the way you look at energy forever.

- Craig Andres, TV News Producer & professional journalist

Chapter 1:

OVERVIEW

"Never underestimate the power of a small dedicated group of people to change the world. Indeed, it is the only thing that ever has."

Margaret Mead

$12,000. Twelve *thousand* dollars. In just a few years, that's how much money you can save on energy costs (gasoline and utility bills). And, you don't have to change your lifestyle or the comfort of your home to save that much or more. Not only can you save enough money to pay for half of a new car, you will also be helping eliminate dependence on foreign oil. Save money, practice financial prudence, and help our nation achieve energy independence. These are just a few of the benefits awaiting you in these pages.

Imagine waking up in the morning, rolling out of bed, and admiring the view of a beautiful spring morning out your bedroom window. Looking up in your back yard, you see your personal wind turbine's black blades rotating silently above the trees. As the sun begins to dart through the trees, you know that soon it will be shining directly on the flexible solar shingles you had installed on the southern-facing roof surfaces. Most of your neighbors still haven't figured out you even have solar at your home since the solar shingles are so subtle they are barely distinguishable from ordinary shingles. While you walk to the bathroom, you smile knowing that today will be another day that you will create more power than you need, and that your local utility will have to pay you for the extra energy you inject into your neighborhood's grid. Too bad it's not much money, but at least it's something. But because you rarely have to pay a utility power bill any more, and in fact get a small check every month, the net impact to your bottom line is over $3,000 per year worth of cost savings and income. Plus, it's comforting knowing that your small renewable energy system means the dirty power plants miles away won't have to burn coal or gas to create your power.

While in the bathroom, you enjoy your luxury shower head's super-low water flow (which still feels just as strong as your old wasteful nozzle) and find it easy to forget that the lighting in the bathroom uses zero electricity due to the sun-tube you installed in all your windowless bathrooms. The rest of your morning is a blur, as you prepare for work on 'auto pilot.' You rarely even think any more about the electricity - and money - you're saving with all your EnergyStar™ appliances (coffee maker, TV, radio, every light bulb in the house, alarm clock, refrigerator, dishwasher, oven, desktop PC, laptop, etc.). But as you enter your garage you snap out of the daze of your morning ritual as you glance at your unusual car.

It always happens this way, in the garage, no matter how busy or distracted you are in the morning: the sight of your car slaps you back into the moment. You walk over and remove the hydrogen gas fueling hose from your super lightweight carbon fiber SUV, admiring the smooth ding-and-scratch-free ivory color panels you selected. Your previous SUV had always gotten blemishes from stray doors banging into it at the company parking lot. Since the hydrogen in your car's safe impact resistant tanks was made from water at your house using wind & solar power, you haven't paid for fuel for two years on either of your vehicles. The savings from that alone is nearly $5,000 per year for both vehicles, and rising fast due to soaring gasoline prices. And because the car is so light and its internal combustion engine burns hydrogen so efficiently, its range is over 400 km (240 miles) per fill. The fuel cell model you have your eye on doubles that range, but its price is still a bit out of your range.

As start the car and pull out of your driveway, you cannot hear a sound from the car since its hybrid electric motor is doing the work at the moment. You know that the only thing that will come out of your tailpipe today is pure water vapor, returning 100% of the water used to make the hydrogen in the first place back into the environment.

Finally, you know that even if your kids forget to turn off the lights in their closets and bathroom – again – you will not have to scold them since the infrared light switch will sense when there is no more movement in the room and turn off the lights for you 10 minutes later.

This example is not science fiction. **This lifestyle is possible today.** All the products, energy savings estimates, and pollution reduction are available right now. The Clean Power Revolution has begun, and within the pages of this book are the answers to show you how to make this lifestyle possible for you and your family today.

But first, we must face reality. We are not well. Even though America is the most prosperous nation on Earth, far more Americans are unhealthy today than what one would expect with our advanced medical facilities and technology. The prevalent sickness and obesity suffered in our nation clearly has many causes. One cause can be quickly identified. We are slowly poisoning our children, and causing premature death among our adult friends, neighbors, and coworkers. Much of this is due in large part to our hopeless and myopic dependence on foreign oil and fossil fuels. The single largest polluter in the United States – by far – is the electric utility industry. "Abt Associates, Inc. (A consulting firm that the EPA has used) released findings in 2004 stating that coal-fired power plants in America are responsible for shortening the lives of over 24,000 people per year (MSNBC);[1] or over 60 people per day. At least one in six women of childbearing age has blood mercury levels high enough to make it unsafe for her to bear children. A 2004 study performed by the CDC (Center for Disease Control) shows that this figure may actually be as high as one in five women. Our nation is now faced with a simple choice: do we embrace clean power or continue down our current path of dependence on fossil fuels? The consequences of either path – in just five years' time – are likely to shock you.

Overwhelming evidence shows that we have already reached peak oil production (meaning we cannot produce more oil than we already do today) and demand continues to grow at breakneck speed by three to four percent each year. At this rate it will take about 40 years to completely run out of all known reserves of oil. Oil prices will skyrocket during the last few decades of production as the cost of extracting these last barrels of oil will become increasingly more expensive – and .the supply will become more and more vulnerable to disruptions from unstable regions of the world.

Uncle Sam has astonishing resource potential to develop clean, renewable, American-made electricity and fuels. In fact, just three states (Kansas, Texas, and North Dakota) have enough wind energy potential to theoretically power the *entire country*. Wind power is the fastest growing source of power in the world, as well as the fastest growing source of power in the United States. However, this statistic is misleading because the wind energy industry, while growing at 35 percent per year, is still quite small relative to the fossil fuel industry. It is easier for an industry to grow quickly when it is small. But the interesting thing to note about the high annual growth rate of wind power is its primary driver: wind is now the lowest cost form of new energy generation. Although the electricity

---

[1] MSNBC Report, June 2004, http://www.msnbc.msn.com/id/5174391/print/1/displaymode/1098/

produced from wind power plants currently costs slightly more per Kilowatt hour (Kwh) than the electricity produced at existing fossil-fired power plants, this will not remain the case as building a new wind-powered electric generation facility costs less than building a power plant that utilizes any other resource.

I hope that this book will serve to stimulate more debate and discussion about the subject of our energy policy. Perhaps with enough readers of this book, our American family will awaken to the reality that this is in fact a problem that will not go away and <u>must</u> be addressed. Therefore, I ask for your attention and a few hours of your time to read this book and make your own decisions based on what you learn. I am confident that you will not be disappointed by investing your time and reading *The Clean Power Revolution*.

This book demonstrates the overwhelming evidence pointing to a proverbial train wreck in our economy if we continue down the path of dependence on fossil fuels. You will learn exactly what the current and future costs will be of this dependence, with every source clearly documented in footnotes for your further review. Then you will learn of a bold new vision for our country called **The Freedom Plan**. This proposed solution, which will be demonstrated to be both realistic and attainable, could perhaps be the single best solution to solve our current energy crisis. The Freedom Plan is based on sound science using technology that is available today. All of the calculations, including the costs of necessary infrastructure, for every objective of The Freedom Plan are realistic. So are the assumptions behind the costs to our economy if we do not act quickly to convert from dirty fossil fuels to clean, renewable power. Although there may be other equally elegant solutions to curb dependence on foreign oil and fossil fuels, The Freedom Plan is the only solution presented for America that can realistically be achieved in 10 years, and completely implemented *without* government assistance. In other words, it is completely *market driven*, as recommended by the August 2004 <u>FORTUNE </u>magazine article, "How to Kick the Oil Habit."

Hence, this book is written for **Republicans** who fear that any major change to our energy policy will cause severe damage to our economy. It will prove just the opposite case: that if we do not take immediate and massive action to curb our dependence on fossil fuels, especially those from overseas, our nation's wealth will be substantially diminished. It will cost us far more if we do not act and, we must act swiftly to prevent additional hemorrhaging in our nation's economy that will only continue to get worse.

This book is also written for **Democrats** who tout the environmental reasons to convert our energy industry to clean power. Admittedly, those are good reasons, but they are not enough. This country is successful due primarily to its diversity and common sense. Those who control the energy industry are not Democrats. This should come as no surprise. Those who control the energy industry are predominantly Republicans. To accomplish anything significant, one must build consensus. Therefore, this book will arm Democrats with the economic reasons for a conversion to clean power to bolster the already compelling health and environmental reasons for such a change.

And this book is written for **all Americans** and to those around the world who are concerned about energy supplies. The subject matter discussed affects all of us. The energy industry is ubiquitous. Its impact is everywhere we turn: in the air we breathe, the water we drink and the food we eat. The prices we pay for everything are affected by the single largest industry in the world, energy. No one can escape the need for energy, and no one can escape the omnipresent news about the pending energy crisis that is now upon us.

Fortunately, a powerful worldwide movement towards clean power already began in 2002 and gained significant momentum in 2004 when energy prices began setting all time record highs. This movement, which I call the **Clean Power Revolution**, is sweeping across the world – but is being driven by grass roots efforts and the sheer will of the people. It is not being initiated by any government, any company, or even by any single demographic or economic trend. Its source of power is a variety of factors, including consumer demand for clean power, fear of dependence on Middle Eastern oil, business demand for diversified energy portfolios, concern over health effects of burning fossil fuels, new evidence proving the human impact on global warming, and anxiety over the mass destruction of the environment as a result of current energy policies. This is the kind of massive movement that comes around only once in a lifetime, cannot be squelched, and will not weaken.

The Clean Power Revolution is here to stay, so what does this mean for America? Will we be left behind in this worldwide effort and cling to our oil-dependent national lifestyle? Will our reluctance to face reality or initiate a major paradigm shift force us to engage our impressive military powers to hang on to fossil fuels for as long as our collective conscience allows us to? Or will we jump at the chance to lead the world in the Clean Power Revolution?

This book will enlighten the reader about the true nature of what is taking place in the energy industry. Readers will have the opportunity to analyze for themselves the economic reasons behind rising energy prices and the expected impact to our economy of maintaining the status quo. More importantly, readers will be able to take a very hard look at how the conversion of our nation to 100 percent clean, renewable, American-made power can **save our nation at least $20 Trillion by 2025** – and exactly where those savings come from. The reader can then study The Freedom Plan and learn how it can be completely implemented in just 10 years at a cost of only **$2 Trillion** with a positive economic impact of **$6 Trillion**. A step by step model is presented showing exactly what steps need to be taken to convert America to 100 percent clean power, using today's technology and today's prices with a realistic timetable and conservative assumptions.

In other words, to completely power 100 percent of our great nation (including the entire electric grid and every car, truck, train and plane in the country) with clean, renewable power, it will cost about the same amount that the United States Government spends every single year in the federal budget (the U.S. government spends about $2 Trillion per year on everything). Then, an economic case is made, in detail, showing how it will cost America a staggering *Ten Times More* than the cost of The Freedom Plan if we don't convert to clean power! Can our nation really afford a true economic cost of $1 Trillion (fully one-half of the entire federal budget) <u>every year</u> in added costs? The facts suggest we cannot wait around for things to somehow improve significantly – they will not, and it is likely to cripple our economy if we don't act soon.

In addition to the educational experience about the energy industry and its pitfalls and warnings, you will have an opportunity near the end of the book to learn how one person truly can make a difference. And you will learn where to look for organizations that support The Freedom Plan.

WHY I WROTE THIS BOOK

The simple answer is: I have dedicated the rest of my life to converting America to clean power. I will do everything in my power to fulfill the promise of The Freedom Plan, unless and until a more elegant solution to this problem presents itself.

The longer answer begins with what originally sparked my interest about clean power. While growing up as a shy child, I always loved to read science magazines such as *Popular Science* and *Omni*. With these as my guide, I learned about solar power and wind energy and asked my parents for solar powered toys for my birthday gifts. Then when my father, a successful, self-made businessman who always taught me to learn things on my own, built his dream home in the suburbs of Kansas City he chose to add a solar shed out back. The large home of approximately 7,500 square feet had four separate furnaces with copper coils running through the air intake portion of each furnace. The solar shed had 30 solar panels which heated a solution of water and anti-freeze. This solution was piped into the home where it circulated through a huge hot water tank (to heat all the hot water needed by the home, the hot tub and the swimming pool) and the furnace coils where it heated the air coming into the furnace. Therefore, the gas burners rarely had to ignite to heat the air in the home, resulting in savings of about **60%** on natural gas heating costs. This was, and still is, unheard for such a large home with four furnaces!

So, I knew renewable energy worked and that it could save money once the investment was made. Then in 1998 while I was running my own successful telecommunications company (35 employees, one of the largest dealers of telecommunications equipment to businesses in the Kansas City area), I read an article about a study performed by a U.S. government lab. This study said that just three states (my home state of Kansas, plus Texas and North Dakota) had enough wind energy potential to power the entire country. I had to read that sentence four times to make sure I read it correctly. I was astounded and I knew instantly that I had found my destiny! I would become a wind energy expert and build wind projects to power the state of Kansas and eventually export extra power to surrounding states. But I was naïve about the energy industry and had many hard lessons to learn.

After starting a wind development business in 1999 and forming an official company entity shortly thereafter, I began leasing land from farmers and ranchers in Kansas who wanted my firm to build wind turbines on their land in exchange for royalties on the energy produced from the wind. That was the easy part because my family name, Helming, is known throughout the Midwest due to my father's success in the agricultural community.

The hard part was getting the local utilities to agree to sign a long term power contract (as described in more detail in Chapter Three) that would allow me to finance the construction of multi-million dollar wind farms. It took me three years to finally get some of those coveted contracts, but by then I knew there was a better way. Eventually I sold my interest in that company in order to form a more exciting company called Krystal Planet that would utilize a unique method of grass roots marketing to finance wind projects without having to rely on utility contracts. In reality, I knew that many utilities would welcome low-cost wind energy if we decided to build a project in their area, but few would be eager to award long-term power contracts for wind energy. Therefore, a market driven approach using consumer and business demand for clean power was the fastest way to build the most wind energy projects.

In early 2003, after having studied how we could convert the entire Midwest (and eventually the nation) to wind power, I realized that there was a large disconnect between the reality of the growing wind energy industry and what the public perception was about wind power. So I knew that educating America about the possibilities with wind energy and related technologies such as hydrogen was paramount to large-scale acceptance of wind farms. Then, in 2004, I saw the signs that our dependence on fossil fuels was beginning to severely weaken our economic strength as a nation. I am extremely proud to be an American and refuse to watch my country fall behind in the race to achieve energy independence. Knowing what I knew about the potential and the cost effectiveness of wind energy, I felt like it was my duty to enlighten as many people as I could about how we could solve the energy crisis now. So, this book is the culmination of three years of research and a solid year of writing whenever I could find the time in between the substantial commitments required of anyone at the 'Helm' of a company growing as fast as Krystal Planet. I hope readers will find this book to be both useful and informative.

THE CLEAN POWER REVOLUTION

Two billion people today live without access to electricity. As their countries continue to industrialize, they will yearn for the creature comforts the developed world already enjoys. Much of these comforts, such as refrigerators, televisions, and automobiles, require electricity or fuel. Where will it come from? What kind of cars will these people be able to drive if oil remains expensive and becomes even more costly? The reader will learn what types of organizations are well positioned to capitalize on what I believe will be the largest shift of wealth the world has ever seen as part of this Clean Power Revolution.

"We are certain that the world's energy needs will increase by 60 percent from now until 2030," says Planet's Voice.[2] Already we have hit peak oil production capacity. It takes about eight years for new oil discoveries to come online and we are no longer finding new sources of oil nearly as fast as we used to. So where will all that extra energy come from?

In 1973, at the time of the first oil embargo, America imported **35 percent** of its oil from overseas. At the time, we all agreed as Americans that we had to reduce

---

[2] Planet's Voice Documents, June 2004, by Louisette Gouverne & Michael Schweres.

our dependence on foreign oil so our economy could never again be held hostage to oil supply interruptions. Yet today, when we should be importing less than 10-20 percent of our oil, how much do you think we import? Distressingly, it is nearly **60 percent**.[3] Every administration and every Congress since 1970 has failed miserably to lead our nation away from dependence on fossil fuels and foreign oil.

Therefore, anyone with any sense knows that the government is <u>not</u> going to make it happen without our help. As Time magazine says, "Americans are heading toward their first major energy crunch since the 1970s. A shortage of natural gas last winter sent home-heating bills spiraling upward. They are expected to keep rising. Higher prices are erasing jobs. As part of a long-standing ritual involving Democrats **and** Republicans, lawmakers and Presidents have devised energy plans that add up to **no** plan at all--not deliberately but by default, leaving the nation with no coherent direction on energy. They've been doing this energy dance off and on for 30 years."[4]

Europe has a significant head-start on converting their continent to clean power. The European Union is larger than the United States (by a full 50 percent) and has a greater landmass. Yet, they have much lower overall wind energy resources. The Sea Wind Europe report shows that off-shore wind farms in Europe could supply enough electricity for every one of the 150 million EU households[5] by 2020. In just 15 years, this ambitious project would create 3 million jobs, add hundreds of millions of Euros to the clean power industry in Europe, and provide electricity produced cheaper than with either nuclear power or coal.

Most of the components for wind generation facilities and related equipment included in The Freedom Plan are "Made in the USA." The technicians needed to install and maintain this equipment will be Americans: our brothers and sisters, dads and moms, and our kids. Once we are able to achieve enough public support as outlined later in this book, this blueprint for energy independence will only take 10 years to complete and could help America lead the world towards a conversion to cost-effective renewable energy production. Shell Oil Company recently stated that "50 percent (½) of all energy will come

---

[3] Time magazine, July 21, 2003 "The U.S. is running out of energy" by Donal L. Barlett & James B. Steele.

[4] TIME Magazine, July 14, 2003.

[5] Sea Wind Europe study undertaken by Garrad Hassan, funded in part by Greenpeace and published in 2004.

from renewable sources by the year 2050."[6] We can choose to lead the world, or be forced by rising energy costs to eventually follow others – and that is not our style.

The time has come to put the nation on a low-carbon diet. We can virtually eliminate carbon dioxide and other greenhouse gas emissions, and save our nation trillions of dollars in the process. Only with a solid economic case can The Freedom Plan truly gain enough momentum to blow through America. This book is intended to arm clean power advocates with the data for such a case and help place the United States firmly back into the lead in the race to be the world leader in the Clean Power Revolution. We can – and should – lead by example by commencing the implementation of The Freedom Plan here, successfully completing it, and then showing every other country in the world how to economically convert to clean, renewable, healthy power using this same model.

Welcome to the revolution!

---

[6] Shell Energy Corporation, Shell Renewables CEO Karen DeSegundo on www.shell.com.

Chapter 2:

THE AUTOMOBILE

Old Ford Roadster courtesy Ford Motor Company

Americans value freedom and independence higher than nearly anything else. Our great nation was founded on those two principles and they remain as important as ever for those born here and for immigrants who arrive here seeking opportunity in our free society.

Perhaps this desire, or need, for freedom is why we have such a love affair with the automobile. Having a car or truck gives us a sense of true freedom: we can drive virtually anywhere, any time in this vast country. Having a car was once a status symbol, a representation of achievement that gave its owner great pride. Today, having a car is, in most areas, a requirement. Just having a vehicle will no longer elicit the admiration you may seek or provide access to the social circles you long for. But having a *nice* car, or a new one, is today still very much a status symbol.

Everyone knows that cars and trucks create pollution. We also know that the smog and ozone in our cities could be improved if we stopped driving as much – or if we drive cleaner cars. But the choices for clean-burning automobiles that are cool, sexy, and affordable are slim to none. Actually there are almost no real options to improve the pollution coming from vehicles. I certainly do not want to stop driving my car, and when buying a new vehicle I want to choose among cars that are aesthetically pleasing and fun to drive. What most people do **not** know, however, is that the engines in our cars were not originally designed to belch such awful pollution into the air we breathe.

The first practical internal combustion engine was invented in 1859 by French engineer Jean Joseph Etienne Lenoir. It ran on Town Gas (also known as illuminating gas; Town Gas is a mixture of primarily hydrogen and carbon monoxide and was used to light street lamps in the early 1900s) and had very low emissions. The four-stroke internal combustion engine was designed by a gentleman named Nicholas Otto in the 19th century. Mr. Otto designed his engine to run on Town Gas. His engines ran quite well on gaseous hydrogen, too. In fact, Otto preferred hydrogen to the waste product of the new oil industry known as gasoline. He felt that gasoline was far too dangerous to use as a fuel![7]

The first diesel engine was designed and built in 1893 by Rudolf Diesel. Mr. Diesel intended his engine to run on vegetable oils and other readily available bio-fuels. At the 1900 World Fair, he ran his engines on peanut oil, which was his fuel of choice for nearly a decade. He made steady improvements to his engine design for years and shared those enhancements with people he trusted. But Mr. Diesel died mysteriously in 1913 while crossing the English Channel on the mail steamer Dresden. His death is still unexplained, although the leading theory is that he was assassinated due to the near immediate rise of a powerful fleet of diesel-engine powered German submarines shortly after his death. Mr. Diesel was friendly to Great Britain, France, and the United States, who were allies against Germany a few years later during World War One.[8]

So although both the gasoline and diesel engines were originally designed and built to run on clean hydrogen or renewable bio-fuels, the adolescent oil industry of the early 1900s discovered that some of their waste products such as gasoline and diesel fuel (created as a byproduct of refining oil for other more valuable products at the time) could be burned in these engines. Since people at the time

---

[7] H2 Nation magazine, December 2003, *"Why Wait for Fuel Cells – Convert the car you drive today"*, by Bob Willis, page 37.
[8] The National Inventors Hall of Fame and www.hempcar.org.

were not thinking about pollution or running out of oil, this seemed like a perfectly reasonable option. Soon thereafter, the political clout held by oil and chemical companies influenced the laws of the U.S. as well as the engine manufacturing and automobile industries to embrace petroleum products as the fuel of choice for all engines. The race to lead the Fossil Fuel Age, and earn billions along the way, had begun.

Now that oil prices are regularly setting new record highs and most oil analysts concur that we have hit peak production (meaning we cannot extract much more oil from the ground – from all sources combined – than we are currently pumping), new fuels are mandatory. It is only a matter of time before oil prices become too high to justify filling SUVs and even fuel efficient vehicles with petroleum products such as gasoline and diesel. This is not a problem for the distant future that we can deal with later. It is already here. As this trend continues in the coming years, greater portions of our income will be required to fill our vehicles, adding further stimulus to the search for alternative fuels.

Fortunately, there are exciting solutions available to us. But first, an explanation of the reasons why we must consider other options is appropriate.

CHALLENGES OF CONTINUED USE OF OIL

High Prices: Oil prices are hovering around record highs in 2004 and are projected to remain high and trend higher over time. The fundamentals of the oil market indicate that a barrel of oil in 2004 should cost about $25. But because two-thirds of the world's oil comes from regions with unstable geo-strategic risks, a risk premium of at least $10 is placed on oil prices.[9] This puts oil at $35 per barrel. But then, when the energy situation in the United States (most notably the rising shortage of natural gas and steadily increasing demand for oil) is accounted for, another $5 per barrel premium is added by the markets. Finally, the growing demand from China and other nations, along with tight supplies, puts at least another $10 premium on each barrel. That is why we are seeing oil at north of $40-50 per barrel and holding steady there. Although the price will likely drop below $50 from time to time, the days of $25 barrels of oil are history.

Competition for limited oil supplies: Thirst for oil from the U.S., China, India and the rest of the world continues to rise, while major new supplies are simply not available. Rising demand creates three problems: higher prices, greater risks of supply disruptions due to minimal excess capacity, and faster depletion of oil

---

[9] From an interview with Philippe Chalmin, economics professor at the Universite Paris-Dauphine, conducted by Planet's Voice in June 2004.

fields. There has not been any new discovery of major oil fields in over 30 years and smaller fields are being discovered much less frequently now than ever before in history.

Peak oil has arrived: Oil will not run out tomorrow, but since we have hit peak production, the cost of extracting oil from the ground will continue to increase along with demand. This places a potent one-two punch on the upward momentum of oil prices.

Pollution: Toxic pollution caused by the oil industry is becoming a significant public issue: smog, ozone, carbon monoxide, particulates from diesel fuel, etc. Public awareness about pollution and demand for cleaner air is at an all-time high with no signs of relenting. Approximately 800 million internal combustion engines burn gasoline or diesel fuel around the world today. Burning a gallon of gas in a vehicle releases a whopping 20 pounds of carbon dioxide into the atmosphere. If you were to capture all that CO_2 (the leading global warming gas) in a large balloon and place it onto a scale, it would actually weigh 20 pounds. But the balloon would have to be quite large, since it would take up 1,220 gallons of air space to hold all the CO_2 created from just one gallon of gas. Imagine the impact to our delicate climate of those 800 million engines around the world burning hundreds of millions of gallons of fuel each week.

Environmental damage: Damage to our fragile environment caused by the oil industry has always been a problem but lately the public has become less tolerant of mishaps. Most people have heard of the Exxon Valdez oil spill in 1989 off the coast of Alaska. But few know that as bad as that oil spill was, it ranks as the twentieth worst oil spill in history. In other words, there have been 19 oil spills that were worse than the Exxon Valdez disaster that spilled 36,000 tons of oil into the ocean. One of those was the tanker Prestige which sank in 2002 off the coast of Spain and spilled 77,000 tons of oil, ranking number 13 on the list. Ruptured gasoline holding tanks and pipelines, oil well fires, and other environmental disasters related to our thirst for oil are equally disturbing.

Fuel additives are required: Fuel additives such as lead, MTBE and others have caused countless health and environmental problems. A widespread source of lead in the global atmosphere is TEL (tetraethyl lead) which was developed in the 1920's by General Motors Corporation and has ultimately caused 90 percent of all of the world's lead pollution.[10] Lead in children causes brain damage, intelligence loss, hypertension, and developmental problems. TEL was marketed as a fuel additive to prevent the cheapest gasoline blends from causing excessive valve-seat recession and piston knock in inexpensive engines. In the 1970's GM

---

[10] Lead poisoning fact sheet from the National Safety Council available at: http://www.nsc.org/library/facts/lead.htm.

acquired catalytic reactor technology and promptly began lobbying for a U.S. phase out of TEL in gasoline. This began in 1975 and was nearly completed by 1986. During that time period, levels of lead in U.S. human blood samples declined by about 78 percent. The European Union finally banned leaded gasoline recently in the year 2000 after seeing the successes of banned leaded gas in the U.S. and Great Britain. Since lead was banned in these major markets, the demand for lower emissions resulted in legislation mandating that oxygen be added to fuel. This was passed in California and other states. Although ethanol (made from corn), methanol or other readily available additives work just fine, the oil industry wanted to develop an additive that could be more easily monopolized to control a new profit center. Hence, MTBE (Methyl Tertiary Butyl Ether) was developed and is proving to be just as poisonous as lead. It is colorless, smells like turpentine and causes immediate discomfort. Studies show that breathing MTBE can cause liver cancer in mice, kidney cancer in rats, and the EPA classifies MTBE as a possible cancer causing carcinogen in humans. By 1993, MTBE had become the 2nd highest organic chemical manufactured in the U.S. This chemical is attracted to water 42 times more readily than air, meaning MTBE finds its way into water supplies, even when the vapor touches water in a rain cloud, raindrop, or moisture in the lining of bodily tissue such as lungs while driving down a busy street.[11] Complaints of eye irritation, headaches, dizziness, burning of the nose and throat, disorientation, and nausea have been traced to the presence of MTBE pollution in Alaska, Montana, Wisconsin, New Jersey, and California so far; many more states are now investigating possible links between health ailments and MTBE exposure.

Danger: Countless children have perished as a result of carbon monoxide poisoning from being left in running cars around the world. Gas tank explosions, car fires caused by spontaneous combustion in fuel supply systems, car fires, and other dangers wreak havoc on lives, families, and insurance companies. All these dangers could be avoided if we were to switch to clean, safe, hydrogen as our primary fuel source.

Health damage: Disease, health ailments, and weakened immune response caused by vehicular pollution are being documented at a record pace. A recent study published in the most respected medical journal in the world, the New England Journal of Medicine, found that "the current levels of air pollution have chronic, adverse effects on lung development in children from the age of 10 to 18 years.[12]" Other studies have linked birth defects in mice to diesel engine pollution. Increasing smog in U.S. cities, caused by nitrogen oxide emissions, has

---

[11] Henry's law Constant (0.024); Solar Hydrogen Civilization by Roy McAllister, p. 52.

[12] New England Journal of Medicine, "The Effect of Air Pollution on Lung Development from 10 to 18 Years of Age" September 9, 2004.

led to the highest incidence of asthma ever seen in U.S. history. From 1980 to 1986, asthma cases nearly doubled to 5.5 percent of the population[13] or 16 million Americans. "Blacks, Hispanics, American Indians and Alaskan natives had higher rates of asthma control problems than whites and Asians;" according to a 2004 Associated Press report. Asthma is so bad in some areas that state attorneys general are filing lawsuits against polluters to force them to clean up their act (see Chapter 3). Officials in nearly 500 counties in the USA in 31 states (representing 159 million people or roughly half the U.S. population) are flunking air quality standards, drawing a federal warning to clean up air pollution.[14] Counties that continue to violate air pollution standards could lose federal highway dollars. Terry Tamminen, the Secretary of the California Environmental Protection Agency, said in a speech in April 2004 that "air pollution sends one of seven children in this region [northern California] to school every day carrying an asthma inhaler. The health of our businesses is also threatened by the rapidly rising prices of fuel, with no end in sight. We cannot build a twentieth century economy on nineteenth century technology. 40 years ago, President Kennedy's bold leadership sent Americans to the moon, using hydrogen and fuel cell technology. Today, we can certainly harness that same technology to take us to work, school, and on a family vacation."

Water supply contamination: One gallon of gasoline creates more than one gallon of water waste when burned; which passes out of the tailpipe and enters the air mostly as toxic water vapor. The water that drips out of tailpipes onto city streets becomes carbonic acid that forms as CO_2 is absorbed into the water coming through the exhaust. As this toxic water enters the sewers and drainage systems around the world, the water supply becomes incrementally poisoned. This polluted water, which did not originate from water (like renewable hydrogen does as shown in Chapter 4) but instead comes from burning a fossil fuel in an engine, adds excess water to the world's oceans that the Earth may not be able to accommodate. Many climate change theories include the impact of water from sources like this (or from melting polar icecaps) added to the ocean's heavier saltwater which could contribute to a slowdown or shutdown of the ocean thermal conveyor system. This effect, which was the premise of the blockbuster movie "The Day After Tomorrow" is based on sound science from numerous supercomputer predictions, could have very severe climate change consequences. This excess water adds up to about 190 million barrels per day.[15] The true extent of this risk is not yet well known.

---

[13] "More in U.S. Suffering from Asthma" The Associated Press, Atlanta, 2004.
[14] "Nearly 500 Counties draw Air Pollution warnings" by John Heilprin," Associated Press, April 16, 2004.
[15] The Solar Hydrogen Civilization by Roy McAllister, p. 173.

Property damage: Vehicle emissions (sulfur dioxide, nitrogen oxide, and other acidic fumes and particulate matter with a low pH) cause property damage such as gradual carbon buildup on buildings and windows (requiring more frequent cleaning and sand blasting), acid deterioration of historic buildings with fragile architectural features, and decay of sensitive objects such as art, outdoor murals, paintings, wooden structures, cloth-covered awnings or table umbrellas, and all things inside a building with its windows open.

THE GOOD NEWS

Fortunately, progress is being made. Hybrids vehicles made by Honda and Toyota were so popular in 2003 and 2004 that most dealers had to stop adding names to their waiting lists. Ford's Hybrid Escape SUV (the world's first hybrid SUV which gets a remarkable 35 MPG in the city) has such low emissions that in many of the most polluted cities the air coming out of the tailpipe is actually cleaner than the air going into the engine. Of the 20,000 Hybrid Escapes slated to be produced in 2004-2005, there are already over 40,000 customers signed up to buy them. Unfortunately, the demand far exceeds the available supply of many of these cleaner, fuel efficient vehicles. Fortunately, this is likely a temporary problem.

Photo of Ford Hybrid Escape SUV courtesy of Ford Motor Company

While the emissions from a hybrid vehicle are significantly lower and cleaner than traditional vehicles, hybrid vehicles still run on gasoline. The name hybrid was coined because of the unique dual-engine nature of these vehicles: they have both a gasoline engine and an electric motor. Hence, the combination of two power sources sharing the propulsion duties lent itself to a namesake of the mixture of two technologies, or a hybrid of both. Usually, the electric motor powers the vehicle during slow speeds and the gasoline engine runs only when needed during acceleration or highway driving. Slowing down uses what are called regenerative brakes, meaning the electric motor runs backward to charge the battery in the vehicle. This battery is the energy source required for the electric motor, but the gasoline engine still requires fuel to operate. The combination effect means the gasoline engine runs less frequently than the more efficient electric motor. This is why hybrids get much better gas mileage and have such low emissions relative to conventional vehicles.

Photo of Toyota Prius sedan courtesy of Toyota Motor Company

HYDROGEN

General Motors has committed over $1 Billion to hydrogen research and development and has over 600 employees working full-time on their hydrogen vehicle program.

Hy-Wire SUV Photo courtesy of General Motors Corporation

Ford, BMW, Toyota, Honda, and most other major manufacturers have also made major commitments to developing hydrogen powered vehicles, spending billions of dollars investing in start-up parts and component suppliers and in their own research programs. Ford has the Model U (run by a clever hydrogen-powered internal combustion engine), BMW has a 7-series luxury sedan program powered by liquid cryogenic hydrogen, and GM's Hy-Wire vehicle and other prototypes have sparked great interest worldwide. Progress is certainly being made, but today there are still no commercially available hydrogen-powered vehicles from a major auto manufacturer. Some small entrepreneurial companies including Krystal Planet (see Chapter 9) do offer hydrogen vehicles options today starting at about $40,000 and many more are certain to be forthcoming.

Photo courtesy of BMW Corporation

Burning safe, gaseous hydrogen, as you will learn in Chapter 4, can actually create negative emissions. This means that when your vehicle's engine burns hydrogen (any internal combustion engine ever built can be converted to run on hydrogen) instead of gasoline or diesel, it actually cleans the air as you drive around. This is because the air going into the combustion chamber of an engine burning hydrogen is heated to 2,200ºC (4,000ºF), so it is steam sterilized very quickly. Anything in the outside air while you are driving around in your smog eating car will also be vaporized. This means you will be cleaning the air of dust, mold spores, pollen, prions, viruses, anthrax, bacteria, particulate matter, airborne hydrocarbons, and pollution from gasoline-powered engines or power plants – just by running errands in your clean, safe, hydrogen-powered, Smog Busting car!

Why Convert an Engine to Burn Hydrogen?

German engineer Rudolph Erren converted over 1,000 vehicles to run on hydrogen, including cars, trucks, and buses. Mr. Erren did these conversions not this year, or last year, or even in the 1990s. He converted all those vehicles in the

1930s! Today in America, there are several reputable companies that can perform a hydrogen conversion on any vehicle. Even Arnold Schwarzenegger has requested a conversion of his Hummer to run on clean hydrogen.

There are six (6) reasons why it makes sense to convert your engine to burn hydrogen instead of gasoline or diesel fuel:[16]

1. **Longer engine life.** Since there is no carbon buildup caused by hydrogen (there is no carbon in hydrogen), your spark plugs and cylinder walls will remain clean and undamaged for much longer. Since there is no sulfur in hydrogen, no corrosive acids will be created to eat away at engine parts.

2. **Negative emissions.** You can actually clean up the air around you as you drive, and help slow global warming and other climate change caused by our current thirst for fossil fuels.

3. **No more oil changes.** The inconvenience and expense of those frequent oil changes can be completely eliminated. Just replace the oil filter (less frequently, especially if you are using a bio-based motor oil) and top off the oil level regularly.

4. **Cold starts are no problem.** Since hydrogen has a lower combustion temperature than gasoline and much lower than diesel, hydrogen powered internal combustion engines will start in the lowest temperatures.

5. **Reduce dependence on foreign oil!** Assuming your hydrogen is produced at home (from your solar or wind powered home hydrogen system as described in later chapters) or from another renewable source, you are no longer contributing to the trade imbalance caused by our helpless dependence on foreign oil. Additionally, you are lowering the need to send our troops to the Middle East to protect our access to oil.

6. **Support the Freedom Plan.** As you will learn in later chapters, the only sure way to guarantee the success of the Clean Power Revolution is for consumers to demand clean, renewable power for their homes and vehicles. Converting your car to hydrogen will stimulate demand for

---

[16] "The Hydrogen Fuel Cell vs. Internal Combustion Engine Debate" by Bob Willis, H2 Nation magazine, December 2003, p. 41.

distributed generation of renewable hydrogen in your area. With steady demand from consumers like you, we can slowly but surely implement the hydrogen economy and take back our country from its dependence on foreign sources of energy and dirty fossil fuels.

Will Hydrogen Become a Reality?

This remains to be seen, but some encouraging signs are beginning to emerge. For example, Governor Arnold Schwarzenegger of California signed executive order S-7-04 on April 20th, 2004 that calls for the completion of a hydrogen transportation network throughout California by the year 2010. This landmark document, signed at the University of California, Davis, will establish a network of hydrogen fueling stations on all California highways – and will forever change the way the world views transportation fuels.

It takes six pounds of gasoline to provide the same energy content as only two pounds of hydrogen.[17] As long as the hydrogen is a) stored in safe, carbon fiber, impact resistant (explosion proof) tanks that are already commercially available and certified by the U.S. Department of Transportation for safe use in vehicles, and b) produced from clean wind or solar power as described in The Freedom Plan, hydrogen can totally eliminate our need for foreign oil. Also it can dramatically improve air and water quality in the process. To learn more about hydrogen powered cars and how you can convert your vehicle to run on hydrogen, join the Clean Power Revolution with any of the companies or non-profit organizations that have endorsed The Freedom Plan (see Chapter 11 for more information).

BIO-FUELS

Bio-Fuels such as ethanol (made from corn), bio-diesel (made primarily from soybeans or waste cooking oil) and other bio-fuels are also seeing steady progress. Any vehicle can run on up to 10 percent ethanol with no problems, and most can operate on up to 25 percent ethanol without incident or damage to the engine.

---

[17] The Solar Hydrogen Civilization by Roy McAllister, p. 57.

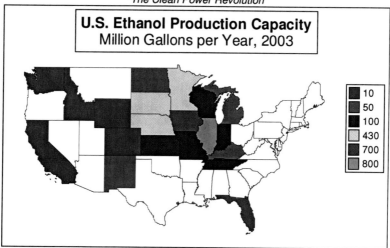

Map courtesy of USDA

Ethanol: Ethanol is a fuel, or fuel additive, made from crops such as corn, sorghum, wheat, sugar cane, sugar beets, molasses, tubers, waste starch and biomass. When burned in an engine, ethanol burns cleaner with fewer emissions, improved fuel economy, less carbon buildup and less sulfuric acid corrosive wear on engine parts than burning gasoline. In 2003, nearly 11 percent of the entire U.S. corn crop was converted into ethanol (see Figure 1 below). After the recent ban on MTBE in California, millions of gallons of ethanol will be added to gasoline on the west coast in order to oxygenate the fuel to mitigate emissions. In 2004, the ethanol industry is projected to produce 3.3 billion gallons of ethanol which should add between 20¢ and 40¢ per bushel to the corn price. Illinois and Iowa have 45 percent of the nation's ethanol production capacity. When all of the new production capacity comes online, eight states will be able to produce at least 100 million gallons of ethanol per year. Minnesota currently has the largest number of ethanol plants, but Iowa is set to take the lead, with four new plants in the planning or construction stages. Combined, the United States has 75 ethanol plants, with another 12 plants underway. In addition to Iowa's four new plants, Illinois is adding two plants; Missouri, South Dakota, and Wisconsin are adding one plant each; and Nebraska has three new plants underway. By 2012, five billion gallons of renewable fuels would make up part of the nation's fuel supply. That is nearly double the 2003 amount of ethanol in use.[18]

---

[18] "Ethanol: Policies, production, and profitability" by Chad E. Hart, Economist, Center for Agriculture and Rural Development, Iowa State University, June 2004.

Ethanol produced today yields about 67 percent more energy than it takes to produce it, including all aspects of the process: growing the corn, harvesting it, transporting it and distilling it into ethanol[19] according to Hosein Shapouri, senior agriculture economist for the U.S. Department of Agriculture. That is an increase from a yield of 34 percent in 2002, just two years prior. Also, ethanol plants produce 2.9 gallons of the fuel per bushel of corn, up from 2.5 gallons just one year prior. Ethanol is also credited with boosting corn prices by 7 to 10 cents per bushel. This adds more fuel to the argument that reducing pollution and converting to clean power will help the U.S. economy, not hurt it.

Another promising benefit of ethanol may be its ability to provide another renewable source for hydrogen. A new process developed at the University of Minnesota's Department of Chemical Engineering and Material Science takes a mixture of ethanol and water, mixes it with air in a fuel injector, and then applies it to a catalyst, Rhodium-Ceria. The result is all the hydrogen is extracted from the ethanol plus some from the water. For about six hours after this major ethanol-to-hydrogen breakthrough was announced, it was the top news story on CNN.com. 140 newspapers around the world picked up the story that day. The chemical reactor used in the process is coincidentally about the size of an ear of corn. With this process of producing hydrogen from corn and using it in fuel cell cars "we can potentially capture 50 percent of the energy stored in corn sugar, whereas converting the sugar to ethanol and burning the ethanol in a car would harvest only about 20 percent of the energy in the corn sugar," says Lanny Schmidt, head of the research team that made the fascinating discovery. Ethanol is also much easier to transport and store than hydrogen, potentially helping to solve one of the early challenges to shifting to hydrogen as a primary transportation fuel. Mr. Schmidt has been studying similar reactions for over 15 years. This is clearly his greatest discovery to date.

Can we actually grow enough ethanol to power the nation? That remains to be seen, but the potential is certainly significant. For example, one acre of corn can produce 300 gallons of ethanol,[20] enough to fuel four cars for one year using a 10 percent ethanol blend. Ethanol clearly can make a difference in reducing our oil consumption and perhaps make a significant difference if higher blends of the corn-based fuel are used such as E85.

---

[19] "Ethanol Revs Up" by Paul Wenske, The Kansas City Star, 9/7/04, pp. D14-15.
[20] Renewable Fuels Association, www.ksgrains.com, reported in the K.C. Star 9/7/04.

Monthly Ethanol and Gasoline Prices, 1999-2003

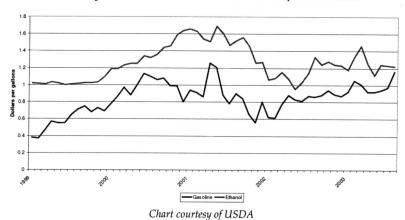

Chart courtesy of USDA

E85: E85 is an alternative fuel produced by blending 85 percent ethanol and 15 percent gasoline. It is a clean-burning, domestically produced, renewable fuel that contributes to decreased dependence on imported oil. Automobiles called flexible fuel vehicles are designed to run on E85. Such vehicles are available from Ford, GM, and many other auto manufacturers. The cost of ethanol, which is the primary driving force behind the cost of E85, has ranged between $1.00 and $1.70 per gallon from 1999 through 2003 (see chart below). Now that gasoline prices are regularly surpassing $2.00 per gallon in 2004, E85 is becoming a very competitive fuel.

Plastics from Corn: Corn can be used to produce ethanol, but it can also replace oil as a renewable source of polymers for *plastics*. When industrial giants Cargill Inc. and Dow Chemical Company launched a plastic made from corn instead of oil in 2002, they thought it would help improve their image among environmentalists. But now that oil prices are soaring, this alternative plastic may become a lucky profit center. "It's a heck of a lot easier to grow a bushel of corn than to find a barrel of oil," says Kathleen Bader, chief executive of Cargill Dow, the joint venture based in Minnetonka, Minnesota that makes the corn-based plastic for the two giant chemical companies.[21] Bio-based plastics can be used in food containers at natural-food stores, in pillows and comforters stuffed with corn-based fiber instead of typical petroleum by products, to-go cups (used by McDonald's Corporation in Europe), fabrics, tennis shoes, and just about anything else that uses plastics typically made from petroleum. Even Sony Corporation is using the product in some models of its Walkman. These bio-

---

[21] The Wall Street Journal, "One Word of Advice" by Thaddeus Herrick, October 2004.

based plastics are expected to remain competitive with oil-based plastics as long as oil prices do not dip below about $25 per barrel.

Biodiesel: Biodiesel can be produced from soybean oil, rape seed oil, yellow grease, animal fats, used cooking oil, peanut oil, and hemp. When it is burned in a diesel engine, it smells like popcorn and has lower emissions than conventional diesel, with the most harmful emissions and particulates being far lower than petroleum-based diesel. The cost of biodiesel depends on the market price for vegetable oil. In general, biodiesel blended at a 20 percent level with petroleum diesel costs approximately 20 cents per gallon more than diesel alone when diesel prices are $1.50 per gallon. In late 2004, as diesel prices consistently hit all-time record high and are over $2 per gallon for the first time in history, biodiesel becomes very attractive.

The premier standard-setting organization in the United States has issued a fuel specification for biodiesel. The American Society of Testing and Materials (ASTM) issued Specification D 6751 for all biodiesel fuel bought and sold in the U.S. in March of 2002, marking a major milestone for the biodiesel industry. Having a full standard in place helps protect consumers from poor products and reduces the cost of buying and selling biodiesel. While many adopted the provisional specification in 1999 (PS 121), those that did not had to negotiate a specification. The final passage of D 6751 streamlines the procurement process for biodiesel for large fleet operators and others.

Biodiesel blends operate in diesel engines, from light to heavy-duty, just like petroleum diesel. **B20** (20percent biodiesel blended with 80percent petroleum-based diesel) works in any diesel engine with few or no modifications to the engine or the fuel system, and provides similar horsepower, torque, and mileage as diesel. Given the other advantages of biodiesel, though, an emission management system with biodiesel is a least-cost alternative. A study by Booz-Allen and Hamilton, Inc., found fleets using a 20 percent biodiesel blend would experience lower total annual costs than other alternative fuels. Similarly, results reported by the University of Georgia indicate biodiesel-powered buses are competitive with other alternatively fueled buses with biodiesel prices as high as $3 per gallon.[22] In late 2004, however, biodiesel prices were actually much lower than that, averaging about $1.65 per gallon.

---

[22] Data found at www.biodiesel.org, based in Jefferson City, Missouri.

NEXT GENERATION CARS

Today's vehicles are made primarily of steel and weigh thousands of pounds. The basic design of the automobile has not changed for nearly 100 years: a steel frame, metal doors and undercarriage, heavy metal engine and transmission, and mechanical linkages to the steering and braking systems. Sure, many improvements have been made over the years, but today's car is not much of a departure from the original Ford Model T. Think about this for a moment: over 80percent of non-commercial miles driven in the U.S. *are in vehicles carrying only one person.* A car is meant to deliver the driver (and perhaps a passenger or two) from one location to another. That is all. But if the average American weighs about 120 pounds (and gradually getting heavier since we rarely walk or bike to our destinations), and the average car weighs 3,000 pounds, then we are building a vehicle with an engine designed to carry 25 times the weight of the person driving it! In other words, we have over-engineered the vehicle with an engine powerful enough to carry around a bunch of steel and metal – the driver's weight is almost an afterthought. Moving far more weight around than what is needed is terribly inefficient. Furthermore, a typical car's internal combustion engine runs at only 15-17 percent efficiency. That means only 17 percent of the energy content of the fuel is actually used to propel the vehicle. So, carrying all that extra weight around with an engine that wastes over 83 percent of the fuel in the form of heat and exhaust is one of the most wasteful activities in the world. Here in America, we do it every day without thinking about it. But it is time we realize there has to be a better way.

Tomorrow's cars will be built using carbon fiber and other composite materials, which are stronger than steel but far lighter. Carbon fiber is used in airplanes, high performance race cars, sailboats and bikes. These carbon fiber parts can be produced using carbon from landfills or even by scrubbing carbon dioxide out of the air, reducing the global warming effect while manufacturing vehicles. Cars using these materials can be much safer than today's vehicles. Even if you have an accident in a heavy steel vehicle, the strength of carbon fiber and crush zone designs that have already been developed, protect occupants from an impact in any direction.

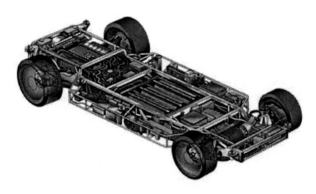

Low profile hydrogen drive-train design, complete with hydrogen storage tanks, fuel cell, electric motor
Image courtesy of General Motors

Body panels can be made from corn-based polymer composites, which are rust-proof, door-ding proof, resist dents and cracks, never require paint (the color is embedded into the panels) and can easily be replaced if the owner tires of the color. Some designs even allow for interchangeable body types, allowing the owner to use the same chassis platform and drive-train but change body panels and interior sections to switch their sports car to a SUV, switch their sedan to a minivan, or anything else. These environmentally-friendly vehicles will no doubt be aerodynamic (low drag resistance) and can have bold, sexy designs unrestrained by the requirements of large engine compartments.

Image courtesy of General Motors

Even the fast-growing racing industry could soon see clean-fueled vehicles racing around the track. The spectators would not have to inhale pollution coming from the racecars, and the noise complaints coming from nearby neighborhoods would be dramatically lower.

Due to the cleaner burning nature of hydrogen powered engines and fuel cells, these carbon fiber hydrogen powered vehicles could have a realistic life of 10-20 years with exceptional warranties to match. This is possible due to the elimination of dozens of repair-prone parts. Hydrogen-powered fuel cell vehicles require no alternator, no transmission (assuming small but powerful electric motors at each of two or four wheels), no brake pads (the motors will quickly slow the vehicle, acting as regenerative brakes to charge the battery whenever the brake pedal is pressed; this allows a lifetime brake warranty), no steering linkages (this function should be electronic even in today's cars but this feature still not widespread), no radiator (depending on the fuel cell), no muffler, no catalytic converter, no oil pump or oil filter, no spark plugs, and no engine. Therefore, trips to the repair shop would be rare, and most diagnostics could be performed electronically by a simple wireless connection built in to the vehicle's onboard computer that connects to your home or office Wi-Fi network.

Such a vehicle could weigh as little as 800 pounds even for a medium-size SUV, requiring a much smaller engine to propel the occupants around. The vehicle could be equipped with either an efficient hydrogen-powered internal combustion engine operating at 40-45 percent efficiency (such engines currently exist and are available from a few select manufacturers) or a hydrogen-powered fuel cell operating at 60-70 percent efficiency. Assuming the latter scenario, such a vehicle could be filled with 8-10 kilograms of hydrogen (equivalent to about 8-10 gallons of gasoline), get 100 MPG, travel up to 1,000 miles without refueling, and emit only pure water. Examples and designs of such vehicles already exist. With enough customer demand, these cars can be mass-produced for $35,000 to $50,000 using today's technology and prices, or less with advances in some of the technologies.

Imagine owning a fuel-cell-powered car made from corn and other biomass renewable polymers that you can fuel with hydrogen made at your own home using only solar power and water. Imagine only having to fuel the vehicle every 1,000 miles, never taking it in to the shop, and knowing it will reliably perform for at least 20 years. That dream will not exist solely in your imagination for long; vehicles like this will be available soon to those who embrace the Clean Power Revolution. Keep reading to learn how you can make this a reality for yourself – and for America.

Chapter 3:

ELECTRICITY –
THE WORST POLLUTER OF
ALL

*"Generating electricity is the single largest source
of air pollution in the United States." MSNBC*

We all know that driving around in vehicles creates pollution. But do you know that generating electricity creates far more pollution than all of our vehicles combined? This is because over 70 percent of our electricity is produced by burning dirty fossil fuels.[23] Over half of our electricity comes from the dirtiest form of power generation: coal-fired power plants. Even brand new coal plants are terrible polluters; there is no such thing as clean coal. The term clean coal is a term used by the industry to refer to the current technology being used in coal plants that has improved the emissions from these plants but it has not made it so that coal-burning electricity plants are clean. Coal plants, even the most modern and 'cleanest' ones, are still the worst polluters by far of the entire electric industry.

According to a recent EPA report, the amount of toxic pollutants in America's air, water and land jumped 5 percent in 2002.[24] This was the biggest increase since the federal government began tracking toxins in 1988. The total was 4.79 billion pounds of poisonous substances spewed into the environment. Mercury and lead – which can damage the developing minds of children – increased by 10 percent and 3 percent respectively. Bill Reilly, EPA administrator for the first President Bush, said "It's not good news." Well, of course not. Electric power

---

[23] Department of Energy,http://www.eia.doe.gov/cneaf/electricity/epa/chapter1.html.
[24] Toxic Release Inventory (TRI) by the U.S. EPA (Environmental Protection Agency), http://www.epa.gov/tri/.

plants increased toxic emissions by 3.5 percent and are responsible for 23 percent of all the nation's worst toxins.

Where does our power come from? This pie chart shows U.S. electric generation by source:

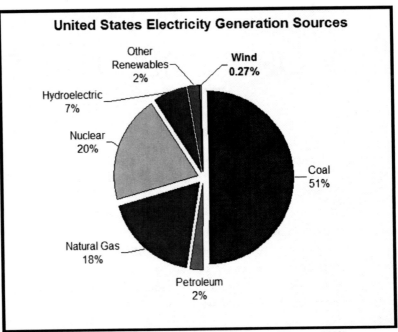

Source: Energy Information Agency (U.S. Dept of Energy, 2002)

A recent study showed that Texans are concerned about cleaner air but are unaware that electricity generation is the leading cause of air pollution.[25] Texas leads the nation in air pollution from coal-fired power plants. Only 5 percent of adults and 2 percent of teens in Texas correctly identified electricity generation as the number one cause of air pollution, yet "cleaner air" was the number one response when given a choice of what to change about the environment.

What emissions spew from power plant smokestacks?

Mercury (Hg): Mercury has been linked to child Autism, Alzheimer's, and heart disease. Mercury is so toxic, just one (1) gram (equivalent to 1/70th of a teaspoon, which would be merely a speck in the palm of your hand) dropped into a 25 acre lake pollutes the entire lake so badly that every fish in the lake becomes unfit to

---

[25] Business Wire, 10/5/04, "TX Leads Nation in Air Pollution from Coal Plants," Austin.

eat.[26] If the concentration gets high enough, mercury can actually kill the fish (see photo). The electric utility industry is responsible for **61 percent** of all airborne mercury in the U.S.,[27] a whopping 298,000 grams every day, 365 days a year. The most advanced emission control technology available today for coal plants can only capture 16 percent of the airborne mercury according to the EPA; the rest goes right up out the smokestack and can travel hundreds of miles in the air and can be dispersed so widely the mercury settles onto millions of acres of land, water, and ecosystems. There are countless numbers of towns sitting downwind from coal plants whose citizens are completely oblivious to the poisons floating down from the air onto their neighborhoods and water supply. Mercury is also known to damage the liver and kidneys and to cause deformities in developing fetuses. Today, the EPA does not even consider mercury a pollutant to be regulated in the electric utility industry, so not a single coal plant in the nation has installed a mercury pollution-control system, nor do any have plans to do so.[28] Many experts claim that since mercury pollution is not even widely monitored at power plants, the estimates based on incomplete data lead many to believe that actual mercury emissions could be much worse than DOE estimates.

Nitrogen oxides (NOx): NOx are a mixture of nitric oxide (NO) and nitrogen dioxide (NO2) and are the leading cause of smog. NOx exacerbate asthma, headaches, lung disease, and create dangerous ground level ozone. NO is colorless, odorless and is oxidized in the atmosphere to form NO2. NO2 is an odorous, brown, acidic, highly-corrosive gas that can affect our health. NOx are critical components of photochemical smog, causing the yellowish-brown color. NOx are harmful to vegetation, can fade and discolor fabrics, reduce visibility, react with surfaces and furnishings, and cause eye, nose, and throat irritation. NOx may cause impaired lung function and increased respiratory infections in young children.[29] Vegetation exposed to high levels of NOx can be identified by damage to foliage, decreased growth or reduced crop yield. Elevated levels of NOx cause damage to the mechanisms that protect the human respiratory tract and can increase a person's susceptibility to, and the severity of, respiratory infections and asthma.[30] From 1980 to 1996, the number of Americans with asthma more than doubled, to almost 15 million, with children under five years

---

[26] 1/70th teaspoon = 0.0703924 mL or .95 grams (Hg temp of 30 ºC, density of 13.5213 grams/mL), source Kimble-Kontes (Vineland, NJ). A 750 MW coal plant emits 400 lbs (181,437 g)/yr of Hg (504 g/day). 600 plants = 108,862,200 g/yr (298,252 g/day).
[27] "Toxic Neighbors" by Thomas E. Natan, Jr. Ph. D, John Stanton, VP National Environmental Trust; Martha Keating, Clean Air Task Force. www.cleartheair.org.
[28] "Benchmarking Air Emissions" published April 2004 by NRDC, CERES, & PSIG, Inc.
[29] EPA Integrated Risk Information System profile for NOx epa.gov/iris/subst/0080.htm.
[30] EPA of Queensland, Australia.

old experiencing the highest rate of increase.[31] Asthma is now the leading cause of missed school days in the Northeast; it has gotten so bad, eight state attorneys general filed a groundbreaking lawsuit in New York against electric utilities to force them to clean up their act.[32] The EPA says long-term exposure to high levels of NOx can cause chronic lung disease and may affect sensory perception by reducing a person's ability to smell an odor or taste food. Electric utilities are responsible for nearly **one-third** of all NOx.[33] In 2002, 29 percent of all power plant boiler capacity had no NOx emissions controls in place or plans to add these controls.[34]

Los Angeles smog photo (8-10-2003). Image available in public domain.

---

[31] Dept Health & Human Services Action Against Asthma, a Strategic Plan, May 2000.

[32] <u>The Record</u>, Hackensack, N.J., "New Jersey joins air pollution suit; 8 states battle 5 key utilities" by Mary Jo Layton, July 21, 2004.

[33] American Lung Association www.lungusa.org; EPA.

[34] Benchmarking Air Emissions" published April 2004 by NRDC, CERES, & PSIG, Inc.

Sulfur Dioxide (SO_2): Sulfur dioxide causes acid rain, acid snow, acid fog or mist, and acid dust. SO_2 pollution causes eye irritation (perhaps permanent eye damage[35]), aggravates heart disease, can cause premature death, decreased pulmonary function, haze, property damage, and can poison ecosystems. The haze formed by SO_2 and NOx affects our enjoyment of national parks, such as the Grand Canyon and the Great Smoky Mountains. Sulfate particles, formed by the reaction of sulfur dioxide (SO_2) from power plants, account for 50 to 70 percent of the visibility reduction in the eastern part of the United States as well as the west.[36] Many power plants use Midwestern and Appalachian coal, some of which contains a lot of sulfur compared to Western coal. The SO_2 and NOx released from power plants rise high into the air and are carried by winds toward the East coast of the U.S. and Canada. When winds blow the acid chemicals into areas where there is wet weather, the acids become part of the rain, snow or fog. In areas where the weather is dry, the acid chemicals may fall to Earth in gases or dusts. Lakes and streams are normally slightly acid, but acid rain can make them very acid. Very acid conditions can damage plant and animal life. Acid lakes and streams have been found all over the country. For instance, lakes in Acadia National Park on Maine's Mt. Desert Island have been very acidic, due to pollution from the Midwest and the East Coast. Streams in Maryland and West Virginia, lakes in the Upper Peninsula of Michigan, and lakes and streams in Florida have also been affected by acid rain. Acid rain had damaged trees in the mountains of Vermont and other states. Red spruce trees at high altitudes appear to be especially sensitive to acid rain. Acid rain can damage health and property as well. SO_2 pollution has been linked to breathing and lung problems in children and in people who have asthma. Even healthy people can have their lungs damaged by acid air pollutants. Acid air pollution can eat away stone buildings and statues.[37] Power plants are responsible for a **63 percent** of all SO_2 emissions in the U.S.[38] In 2002, a whopping 69 percent of all power plant boiler capacity had no SO_2 emissions controls installed or planned.[39]

Carbon dioxide (CO_2): Carbon dioxide is the leading global warming gas. The United States emits over 1.5 billion metric tons of carbon every year.[40] The amount of carbon being spewed into the atmosphere every year is equivalent in weight to 3,000 buildings the size of either of the World Trade Center towers

[35] AERIAS website: http://www.aerias.org/kview.asp?DocId=143&spaceid=1&subid=7.

[36] EPA website: http://www.epa.gov/airmarkets/reghaze/index.html.

[37] EPA websites, including http://www.epa.gov/air/urbanair/so2/hlth1.html.

[38] "Benchmarking Air Emissions" published April 2004 by NRDC, CERES, & PSIG, Inc.

[39] "Benchmarking Air Emissions" published April 2004 by NRDC, CERES, & PSIG, Inc.

[40] DOE reports & websites including:http://cdiac.esd.ornl.gov/trends/emis/tre_usa.htm, National CO2 Emissions from Fossil-Fuel Burning, Cement Manufacture & Gas Flaring, August 28, 2003, CO2 Info Analysis Center, Oak Ridge National Laboratory.

(they weighed about 500,000 tons each) or 3.5 million Boeing 747 jetliners every year (a Boeing 747 weighs 836,000 pounds). This amount of CO_2 pollution is equal to more than twice the amount of the next largest CO_2 polluter, China. With only 5 percent of the world's population living in the U.S., I am embarrassed to admit that my country emits over one- third (36 percent) of all of the world's CO_2. We are by far the world's worst contributor to global warming and climate change. Most people today agree that global warming is a reality; anyone who does not is simply uninformed or defiant. The signs are all around us (see National Geographic section below). There is still some debate, however, over how much impact humans are making to global warming. Although many questions remain unanswered, the sheer magnitude of the scientific research performed in the last decade have proven beyond a shadow of a doubt that mankind is indeed a major, perhaps the primary, cause of global warming. In Chapter 6 you will learn about the heavy economic toll climate change is wreaking on America. The environmental effects are shocking. An entire island chain in the south Pacific, Tuvalu (the world's fourth-smallest country), is at risk of becoming the first nation to be completely submerged due to rising ocean waters;[41] a U.S. Department of Defense study, typically conservative, says that by 2020 climate change will surpass terrorism as the greatest threat to national security.[42] The hit Hollywood movie "The Day After Tomorrow" released in 2004 by 20th Century Fox, while overdone with typical Hollywood drama, was based on real science from the 10-year-long World Ocean Circulation Experiment launched in 1990 which suggests the ocean conveyor belt could slow, then eventually shut down as freshwater (which is lighter than saltwater) continues to flow faster than it ever has in history into oceans from melting ice.[43] Global warming is predicted to bring more violent weather (including hurricanes and typhoons as ocean waters warm) costing nations billions in damage to property as well as lost lives, change crop output, reduce freshwater availability, lower fish and game supplies, wipe out entire species of animals, insects, and plant life, destroy coral reefs, melt the last remaining glaciers, cause Siberian-like climate in Great Britain, Scandinavia, and the American Northeast, flood coastal cities beyond repair, and induce widespread wars over ever more limited supplies of food and water. There are credible theories showing that as ocean temperatures rise, methane hydrates (there are enormous amounts of methane stored safely in these hydrate crystals[44]) that are locked in cold, deep coastal shelf regions will

---

[41] "Tuvalu Toodle Loo" by Mark Levine, Outside Magazine, December 2002.

[42] "An Abrupt Climate Change Scenario and Its Implications for United States National Security" commissioned by the U.S. Department of Defense's Office of Net Assessment, reported by David Stipp in FORTUNE magazine, January 26, 2004.

[43] National Geographic magazine, "Now What?" by Virginia Morell, pp. 27 & 66.

[44] U.S. Geological Survey: http://marine.usgs.gov/fact-sheets/gas-hydrates/title.html.

begin melting, sending methane gas bubbling to the surface in huge quantities. This is of grave concern because methane is 27 times more potent as a global warming gas than CO_2. Therefore, the release of this methane due to the global warming effects of CO_2 emissions could accelerate climate change at a geometrically faster rate than ever before predicted. The challenge with these predictions is that the Earth is so complex a system, no computer model can accurately predict with 100 percent certainty what the outcome will really be. We are performing a grand, dangerous experiment on a global scale – and the subjects of that experiment, mankind and our fellow creatures on Earth, have no where else to go if the experiment goes badly. Power plants are responsible for **39 percent** of all CO_2 emissions[45] in the United States.

Antarctica Melting Ice Shelf – courtesy London Times

Solid particulate matter: These particles induce asthma attacks and contribute to upper respiratory illness, heart attacks and lung disease. Recent studies have shown that 23,600 Americans die prematurely every year as a direct result of power plant pollution (including particulates).[46] That means 64 Americans die every day from asthma attacks, lung disease, heart attacks and upper respiratory failure. Imagine waking up every morning to watch on the news that another 64 people died from power plants. Would we tolerate that for long? Of course not, but because the message is not getting through via the mainstream media, we do not yet witness a widespread Clean Power Revolution. But, as you will learn in Chapters 9 and 10, this revolution has already begun and is growing more powerful every day. I encourage you to join this cause in your own way if you feel it is appropriate. There are likely several organizations operating in your area to choose from that have officially endorsed The Freedom Plan. U.S. power plants are killing Americans by emitting over one million short tons of PM_{10}

---

[45] "Benchmarking Air Emissions" published April 2004 by NRDC, CERES, & PSIG, Inc.

[46] The Washington Post, "How Power Plants Kill" June 26, 2004, p. A-21.

particulates (those greater than 10 microns in diameter) per year plus another 700,000 short tons of PM$_{2.5}$ particulates per year (those between 2.5 and 10 microns in diameter).[47] Breathing this soot from power plants, factories and streets cause more than just respiratory damage. Until 2004, there had been little evidence that any air pollutant could cause genetic damage that could be passed on to future generations. In 2002 Canadian scientists housed mice downwind from plants spewing particulate emissions and tested their offspring. The male mice passed on double the DNA mutations as mice living in the countryside.[48] This is the first direct link ever found between particulates and genetic mutation. Particulate pollution is also linked to bronchitis, emphysema, lung cancer, heart disease, strokes and birth defects.[49]

Heavy metals: In addition to the devastating effect mercury has on the human nervous system, power plants spew thousands of pounds of other heavy metals into the air such as **lead** at 200,000 pounds per year (particularly harmful to children and can cause brain damage, impair growth, damage kidneys and cause learning and behavioral problems), **cadmium** (a cancer-causing toxin), **arsenic** at 113,000 pounds per year (poisonous in any quantity, can be deadly with as little as 20 micrograms[50]), **chromium** at 276,000 pounds per year (a carcinogen known to cause many forms of cancer documented in the Universal Studios film "Erin Brockovitch" which was based on a true story of over 600 Hinkley, Califronia residents who were severely poisoned by chromium; many died), and many other heavy metals in smaller quantities. In addition to air pollution, the electric utility industry dumps these toxins directly onto land and into water supplies near the plant. In many cases, the amount of heavy metals dumped into our environment via land and water far exceeds the amount belched out as air pollution. For example, an additional 23 million pounds of lead is dumped into our waterways and onto land every year from the mining and burning of coal for electricity generation.[51] Other toxins released by power plants include (this is a partial listing): dioxin, hydrochloric acid, sulfuric acid, ammonia, formaldehyde, trace uranium, chlorine, volatile organic compounds, poisonous carbon monoxide (which can cause migraine headaches and even death) and dozens more with names far too difficult to pronounce.

---

[47] EPA website: http://www.epa.gov/airtrends/pm2.html.
[48] SCIENCE Journal, study by McMaster University, Ontario, Canada, May 15, 2004.
[49] The Pitch "Damage Control" by Nadia Pflaum, July 8-14, 2004.
[50] http://www.dchtrust.org/arsenic%20king%20of%20poison.htm.
[51] "Toxic Neighbors" by Thomas E. Natan, Jr. Ph. D, John Stanton, VP National Environmental Trust; Martha Keating, Clean Air Task Force. www.cleartheair.org.

The EPA states that power plant pollution is by far the largest contributor to air pollution in the northeastern United States. Many state departments of health and human services have posted public warnings on their websites and near lakes and rivers warning women of child-bearing age and children that they should not eat too much of certain types of fish – and some fish they should not eat at all.

National Public Radio (NPR) commentator Richard Harris reported on June 9, 2004 that measurements taken in Pennsylvania during the August 14th, 2003 electricity blackouts across the northeastern United States showed a significant drop in air pollution. Russel Dickerson at the University of Maryland leads a team that flies a research aircraft that takes air measurements in the region. They decided to fly the plane over the area suffering from the blackout to determine what the impact would be of all the power plants being shut down for a few days. First, they took air samples in Maryland and Virginia, which still had power, and then they flew north to Pennsylvania, in the heart of the affected region. The air in that area is usually choked with sulfur dioxide, ozone, and other pollution. "The sky was bluer, the visibility increased by an average of 20 miles, the concentrations of ozone fell by 50 percent and by a factor of five [500 percent reduction] for sulfur dioxide," says Dickerson. "It was remarkable how quickly the air quality improved when the power plants went offline."

Another byproduct of burning fossil fuels is the creation of ground-level ozone, which is known to be harmful to plants and animals.[52] The NY Times says, "In the stratosphere, ozone forms a layer that protects life on earth from ultraviolet radiation; that layer has been decreasing, while ozone concentrations in the air that people breath have been increasing...It is known that pollution in urban areas could produce larger ozone concentrations downwind...Cities do have more kinds of pollution than rural areas, and some of these pollutants are the chemical precursors of ozone, like nitrogen oxides. Sunlight acts on these substances, initiating chemical reactions that produce ozone." Hence, the more pollution is created during this Fossil Fuel Age, the worse the damaging side effects become.

COAL PLANTS – More dangers than most people realize

What was the worst environmental disaster in America? If you guessed the Exxon Valdez oil spill, you would be wrong. October of the year 2000 was when a monstrous environmental disaster occurred that was **25 *times worse*** than

---

[52] The New York Times, July 10, 2003, "City Trees Outgrow Rural Cousins, Study Credits Urban Chemistry" by James Gorman, p. B1-B5.

Exxon Valdez. The story begins in Appalachia, where a dedicated federal employee by the name of Jack Spadaro had worked as the head of the National Mine Safety and Health Academy (MSHA). MSHA, a branch of the Department of Labor, trains mining inspectors. This region has seen extensive coal mining for decades. CBS News correspondent Bob Simon covered the story on April 4th, 2004[53] when he learned that Mr. Spadaro had been fired by MSHA after blowing the whistle on an attempted cover-up of the worst environmental disaster even seen in the eastern United States. "The Bush administration came in and the scope of our investigation was considerably shortened. I had never seen something so corrupt and lawless in my entire career," said Spadaro. "I've been in government since Richard Nixon. I've been through the Reagan administration, Carter and Clinton. I've never seen anything like this."

What he's talking about is a government cover-up of an investigation into the disaster in October of 2000 when 300 million gallons of coal slurry - thick pudding-like waste from mining operations - flooded land, polluted rivers and destroyed property in Eastern Kentucky and West Virginia. The slurry contained toxic poisons produced as a by-product of coal mining such as arsenic and mercury. **"It polluted 100 miles of stream, killed everything in the streams, all the way to the Ohio River,"** says Spadaro, who was second in command of the team investigating the accident at the time. The slurry had been contained in an enormous reservoir, called an impoundment, which is owned by the Massey Energy Company. One night, the heavy liquid broke through the bottom of the reservoir, flooded the abandoned coalmines below it and roared out into the streams.

During the investigation, it was discovered that there had been a previous spill in 1994 at the same impoundment. The mining company claimed it had taken measures to make sure it wouldn't happen again, but an engineer working for the company said the problem had not been fixed, and that both he and the company knew another spill was virtually inevitable. "We knew there would be another breakthrough,'" says Spadaro. "We knew. And I asked him how many people in the company knew and he said, 'Well, at least five people.'"

So why didn't they fix it? "It would have been expensive to find another site. And I think they were willing to take the risk … It was a certainty," says Spadaro. He says it was a certainty because there was only a very thin layer of rock at the bottom of the reservoir. But that's not what the mining company had told the government. "They told the government that there was a solid coal

---

[53] CBS: http://www.cbsnews.com/stories/2004/04/01/60minutes/main609889.shtml.

barrier, at least 70 to 80 feet wide between the mine workings and the bottom of the reservoir," says Spadaro of the barrier, which is less than 20 feet. "They were misrepresenting the facts ... and they knew that. The company knew that and I'm sorry to say I believe some people within the government knew that."

"This was a catastrophic failure. By the grace of God only did we avoid fatalities," says McAteer, who was Spadaro's boss at the time. The investigators were going to cite the coal company for serious violations that would probably have led to large fines and even criminal charges. But all that changed when the Bush administration took over and decided that the country needed more energy – and less regulation of energy companies. The investigation into Massey Energy, a generous contributor to the Republican Party, was cut short. "The Bush administration came in and the scope of our investigation was considerably shortened, and we were told to wrap it up in a few weeks," says Spadaro. Because he refused to sign a watered down report quickly pushed through by the Bush Administration, lowering the violations from eight down to two minor infractions and lowering the fines to a paltry $110,000, Spadaro was fired.

"They cut it off. They did," says Ellen Smith, who publishes the country's only newsletter devoted entirely to mine safety and health. She's been writing about the mining industry for 16 years. The story was also covered by **60 Minutes**, which interviewed Spadaro but was declined an interview by Massey Energy, the new head of MSHA Dave Lauriski, and his boss Elaine Chao, the secretary of labor.

There are hundreds of impoundments like this one throughout the Appalachian Mountains, and more exist in many other parts of the country, wherever coal has been mined. Similar slurries are found near the 600 coal-fired power plants in the United States, where the sludge from toxic coal ash is stored onsite in huge landfills. Evidence is now showing these landfills are seeping dangerous chemicals and heavy metals into our groundwater. It is only a matter of time before another impoundment collapses and floods our precious waterways with more poisons, eventually spilling millions of dead organisms and toxic chemicals into our already overstressed oceans and fragile coastal ecosystems.

The Great Energy Scam
A TIME magazine investigation found that a select group of insiders are lining their pocketbooks at the expense of the American taxpayer to the tune of $1 Billion per year in a shocking article called "The Great Energy Scam."[54] The

---

[54] TIME "The Great Energy Scam" by Donald L. Barlett & James B. Steele 10/13/03

authors write, "What you would see behind the curtain is a scheme that would make the Wizard of Oz envious. And you wouldn't be amused, because as an American taxpayer, you're paying for it. About a mile off the twisting, two-lane road to the south of Central City, Pennsylvania, set back in the woods along a private road, past the truck scales and the raw-coal stockpile, invisible from the highway, is the Shade Creek processing plant of PBS Coals Inc. There freshly mined coal is washed, the sulfur, rock, ash and other impurities removed and the cleaned coal carried by an overhead conveyor belt across the dusty road. It goes into a building on the other side that is operated by a second company, Central City Synfuels. Another belt comes out of that building – off limits to the public – carrying what looks a lot like the same coal back across the road and dumps it on a stockpile. Then it is loaded into railcars and shipped to electric utilities. Except it isn't coal any longer. Forget that it looks like coal. And will burn like coal. It is now called "synthetic fuel." As such, the coal-like product, along with roughly 50 million tons of similar stuff from more than 50 similar plants in Pennsylvania, West Virginia, Alabama and other states, is worth more than $1 billion a year in federal income-tax credits, a corporate giveaway protected by a bipartisan group of supporters in Congress."

The article continues, "Those who have profited from the system range from fast-buck artists to giant corporations. They include one of the nation's largest hotel operators, a commodities trader barred from the industry for fraudulent practices, a chain of electronics stores, an electric utility that unplugged the lights during the great blackout of 2003, technology firms run by friends of influential lawmakers, limited partnerships of wealthy investors and scores of individuals and businesses preferring to keep their identities secret. To qualify for the tax credits, the makers of this so-called synfuel do not have to prove that they are making a better kind of coal, one that burns more efficiently or offers any other benefit. By IRS ruling, they need only modify the chemical composition of coal. As a result, dozens of plants have sprung up across America to carry out a process that in many cases is so slight that critics call it **spray and pray**, a reference to their hopes that no one will peek too closely. "You can't believe what goes on," a government official long involved with the coal industry told TIME, blaming Congress for its role in perpetuating the handout. "The people who spend the tax money don't have a clue." The IRS launched an investigation last June into the "scientific validity of the test procedures" used to measure compliance with the minimal standards, but the synfuels credit has enough support in Congress that [supporters of the subsidy] have tried to block the IRS probe."

This is just one more sign that government efforts to curb imports of foreign energy frequently results in waste, corruption and little or no impact on reducing fossil fuel use.

NATIONAL GEOGRAPHIC – Courageously Telling the Truth

The September 2004 issue of National Geographic magazine has 73 pages of full color photographs and coverage of the perilous effects of global warming in its cover story "Global Warning – Bulletins from a Warmer World." It is easily the most comprehensive and powerful coverage ever provided on the subject by a magazine. I encourage you to contact National Geographic and request a copy (1-800-647-5463 or www.nationalgeographic.com). Key excerpts and stories are paraphrased below.

MELTING ICE, RISING SEAS

When President Taft created Glacier National Park in 1910, it was home to an estimated 150 glaciers. Now it contains fewer than 30 and most of those remaining have shrunk in area by two-thirds. By 2035, all of them will have completely vanished predicts Daniel Fagre of the U.S. Geological Survey Global Change Research Program. Peru's Quelccaya ice cap is melting by 600 feet per year and will disappear by 2100, leaving those who rely on it for drinking water and electricity thirsty and powerless. The famed snows of Kilimanjaro are expected to vanish within 15 years. Since 1978 the Arctic sea ice near Canada has decreased by 9 percent per decade. Scientists using sonar data from submarines document a 40 percent thinning of what ice remains since 1978, and some predict it could be gone completely by 2100. Already, a channel of ice-free water called the Northwest Passage is open much of the year north of Canada. The Passage connects the Atlantic and Pacific oceans in a route 7,000 kilometers shorter than the route through the Panama Canal, allowing marine vessels to cut shipping times significantly. By some estimates, the passage will be open year-round within a decade.[55] Chunks of ice are breaking off glaciers in record numbers and size. For example, a chunk of ice nearly as large as Rhode Island (1,250 square miles) that was part of the Larsen Ice Shelf in Antarctica collapsed in 2002 in record time: it took just over a month to break off and begin melting. Without ice shelves to act as dams, glaciers could migrate faster toward the ocean and contribute further to rising sea levels.

Today, the sea level is rising one inch every decade but the rate of rise appears to be accelerating. The Louisiana gulf coast has already seen steady evacuations of towns, neighborhoods and tourist destinations. This is because the southern

---

[55] The Economist, "Breaking The Ice" August 21-27, 2004.

Louisiana coasts are literally sinking by roughly three feet per century (called subsidence) while Gulf of Mexico waters continue to rise. This powerful combination has left homes soaking in cemeteries under water, and former coastline is now a feeding ground for marine life. More than a hundred million people worldwide live within three feet of average sea level. Large Cities near coastal plains or river deltas such as New York, Miami, Los Angeles, Shanghai, Bangkok, Jakarta, Tokyo, Rio de Janeiro, and more are at risk. In the Netherlands, half of the landmass is at or below sea level. Every inch of sea level rise results in eight feet of horizontal retreat of shorelines from erosion, calculates Bruce Douglas, a coastal researcher at Florida International University (FIU).

Impact of rising sea levels:

- A mere four inches would completely flood many low-lying South Sea islands.

- 1.5 feet would swallow 75 percent of coastal Louisiana wetlands.

- 3 feet of rising currents would displace 70 million people in Bangladesh alone.

The two largest ice sheets in the world are in Greenland and Antarctica. Records from mud forams on the Irish Sea coast show that there was a rapid 35-foot rise in global sea level 19,000 years ago. "Two ice sheets the size of Greenland's today must have melted [after that ice age ended] in just a few hundred years" says Peter Clark of Oregon State University. If the West Antarctic ice shelf were to break up and melt, it alone contains enough ice to raise the sea level by 20 feet. Even without the largest ice sheets melting, the IPCC (Intergovernmental Panel on Climate Change) projects that by 2100, the sea levels could rise up to 35 inches which would be "an unmitigated disaster," according to Bruce Douglas of FIU.

If a portion of the massive amounts of ocean methane hydrates were to begin bubbling methane gas to the surface as described earlier in this chapter, it could cause "runaway" global warming on a mass scale due to the potent impact of methane to the warming cycle. If this were to occur, all the ice in the Artic, Antarctica, Greenland and Alaska could melt and, as a result, the sea level could rise 100 feet or more.[56] Recent geological evidence shows that rapid deglaciation after ancient glacial periods was accelerated by this methane mechanism.[57]

---

[56] The Capital Times "Runaway Global Warming from Oceans' Methane is Real Threat" 9/4/2003.
[57] SCIENCE Journal ""Historic Global Warming linked to Methane Release" 11/19/99.

Water temperatures are rising in all ocean basins and are increasing at much deeper depths than previously thought, according to NOAA (National Oceanic and Atmospheric Administration). Perhaps the largest oceanic change ever recorded is the declining salinity of the sub-polar seas near the North Atlantic. This could have considerable effects on the ocean conveyor belt. Robert Gagosian, president of the Woods Hole Oceanographic Institution, believes our oceans are the key to a potential global climate change disaster. If enough change occurs in ocean temperature and salinity, disruption to the North Atlantic thermohaline circulation could slow or even halt the conveyor belt – causing drastic climate changes in just one decade. This and other frightening studies were referenced in a study ordered and paid for by the Pentagon predicting that drastic climate change could surpass terrorism as the greatest threat to our nation. In fact, the worst case scenario of the study warns that famine, wars and general chaos could be triggered by sudden climate change in the next 15 years.[58] The authors say that the possible scenarios would challenge national security in ways that should be considered immediately. The U.S. military is already addressing this concern by adding four enormous wind turbines to the U.S. Navy base in Cuba at Guantanamo Bay to cut costs and pollution in the Caribbean. The $11.6 million, 3.8 megawatt project is expected to cut diesel fuel costs by 25 percent and reduce pollution and greenhouse gas emissions by 13 million pounds per year.[59]

"We have created the environment in which our children and grandchildren are going to live," says Tim Barnett of the Scripps Institution of Oceanography. We owe it to them to prepare for higher temperatures and changed weather – and to avoid compounding the damage. "We've got to do it" adds Jerry Mahlman of the National Center for Atmospheric Research.

ALASKA

Alaska is like a looking glass into our future, since the effects of global warming are felt first in northern latitudes. "If you want to see what will be happening in the rest of the world 25 years from now, just look at what's happening in the Arctic," said Robert Corell, a former top National Science Foundation scientist. "Alaska is the melting tip of the iceberg, the panting canary" said Deborah Williams, a former chief Interior Department official during the Clinton administration and current executive director of the Alaska Conservation

---

[58] Knight Ridder Newspapers "New Weather Patterns Could Stir Turmoil" by Seth Borenstein, February 29, 2004, report available on www.ems.org/climate.
[59] The Miami Herald "Towering Windmills to cut costs for the Navy" by Carol Resenberg, November 21, 2004.

Foundation. Snow and ice have an albedo effect (they reflect a lot of solar energy). So, as the snow, ice, and permafrost melt worldwide and expose the darker land beneath, the warming trend accelerates even faster. An estimated 25 cubic miles of the Buckskin Glacier in Denali National Park and Preserve melts into the ocean every year, which is the single largest contributor to sea-level rise on Earth.

Alaska's average winter temperature has warmed by 8 degrees since the 1960's, and 2002 was the hottest year in Alaskan history.[60] Near Portage Glacier, a top tourist attraction, the $8 million Begich-Boggs visitor center was built in 1986. By 1993, the glacier had receded so much that tourists can no longer even see the glacier from the visitor's center. Prudhoe Bay permafrost, which supports the only roads to the area that are relied upon by oil drilling operators and hunters, was strong enough to support vehicular traffic for 200 days a year in 1970. Today, workers can only drive on the permafrost for 103 days, requiring year-long stockpiling of supplies and extended lodging. The town of Northway loses parts of its highway into sinkholes 25 feet deep, other highways are collapsing, towns have closed down and people are being forced to move out of the area since their homes fall over when the permafrost cannot support the foundations any longer.

Melting of Portage Glacier, Alaska

Spruce bark beetles are doing twice the damage to the vast spruce forests than they used to a mere 20 years ago since the warmer temperatures allow more of the larvae to survive. Four million acres of Alaska spruce are now dead. "We

---

[60] "In Alaska, a preview of global warming" by Seth Borenstein, KC Star, July 21, 2003.

went into overdrive, climate-wise" says ecologist Ed Berg of the warmer summers that began in 1987. Warm water parasites are now present in the lower Yukon River and are attacking king salmon and herring, significantly reducing the fish populations. The Bristol Bay salmon spawning runs are disrupted by warmer water as well as new warm-water parasites.

Alaskan glaciers add 13.2 trillion gallons (equivalent to 13 million Olympic-sized swimming pools) of melted water to the ocean each year. The rate of water runoff has doubled in just a few decades, according to studies performed with airborne lasers by University of Alaska-Fairbanks scientists. The Columbia Glacier sends dangerous icebergs into Prince William Sound shipping lanes. Alaska's melting glaciers are currently the number one reason the world's sea level is rising.

DROUGHT

Another dangerous impact of global warming is drought. Lake Chad, once among the largest lakes on the African continent, has withered by 90 percent since 1960. Fish populations have been decimated and water available for farm irrigation is nearly nonexistent in this region. Lake Powell, in Utah, is only half full after five years of drought in the western United States. The hydroelectricity and water provided by this lake to millions in the southwestern U.S. are at risk. Many mountain streams in California's Sierra Nevada region are now completely dry by summer. Some avian experts suspect that droughts are contributing to a rise in the West Nile virus and other viruses due to a concentration of migratory birds and virus-carrying mosquitoes at remaining water sources where diseases spread more quickly. If droughts increase in severity as global warming continues – as many scientists predict they will – drastic water shortages could arise in heavily populated desert areas in Arizona, New Mexico, Nevada, California, and Utah, as well as dozens of other regions all over the world. This could lead to more regional conflicts as nations and tribes fight over declining water and food supplies.

THE DEBATE ON GLOBAL WARMING

Is it possible to once and for all silence the skeptics who continue to insist that global warming is not real? Yes, that is easy. What is considerably more difficult is proving exactly what is causing global warming, and how much impact mankind has on causing it. This section will attempt to succinctly address each of these issues by providing a wealth of facts and noting a number of recent studies that have employed the most sophisticated technology available today.

As the three primary greenhouse gases (CO_2, NOx and methane) from electricity generation and fossil fuel combustion in engines continue to spew into our atmosphere, the correlation of rising temperatures to rising emissions eerily mirror one another. Charting the three gases alongside the global temperature shows nearly identical patterns: they remain basically flat until the mid 1800s, then all three trend upward together until 1950 when they turn sharply upward. "We have very significantly changed the atmospheric concentration of these gases. We know their radiative properties," says Peter Tans of NOAA's Climate Monitoring and Diagnostics Laboratory in Boulder, Colorado. "It is inconceivable to me that the increase would not have a significant impact on climate."

Is it possible to accurately measure historical levels of global temperatures and gases such as carbon dioxide, methane, and nitrogen oxide that are present in the atmosphere? Yes, absolutely. In fact, there are multiple methods that scientists employ to measure such trends. Every single method draws the exact same conclusion: mankind is significantly altering the global climate. Other factors play a part, such as dust and gases from volcanic eruptions, natural process such as methane emissions from animal waste and other organisms, and to a lesser degree sun flares (although the Earth's many layers of atmosphere blocks nearly all of the non-magnetic radiation emanating from the Sun). But, those non-human factors can be readily quantified over the last 150 years, allowing science to easily distinguish between human and natural sources of global warming gases.

No one can question whether global warming is occurring after a short glance at the following chart from NASA:[61]

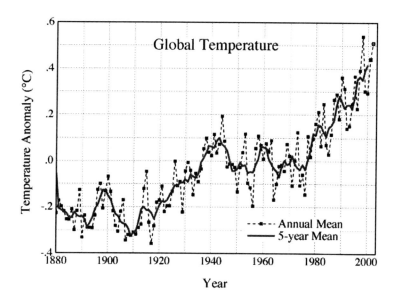

Critics of global warming theories and computer models have relied on a study performed by scientists at the University of Alabama in Huntsville. Incidentally, Alabama is home to some of the largest oil and gas industry plants and chemical factories that rely on petroleum products for raw materials. Over a decade ago, these scientists analyzed satellite temperature data and found that global temperatures were flat – and did not show a rising trend. That one study has been cited again and again by all those who dispute the fact that Earth is getting warmer. But there are several problems with relying on this lone study. One, satellites have not been measuring global temperatures for a long enough period of time, so the data used by that study was only a snapshot of a couple decades' worth of measurements. Two, satellites do not directly measure temperatures but look instead at microwaves emitted by oxygen molecules scattered fairly evenly throughout the atmosphere. But oxygen levels have been steadily decreasing as more CO_2 is released, which lowers the total amount of microwaves emitted. Hence, satellite measurements should show a *decline* in global temperature with fewer microwaves coming from fewer oxygen molecules. Since that was not the case (the University of Alabama study found a fairly stable level of microwaves), it suggests temperatures must actually be

---

[61] http://www.giss.nasa.gov/research/observe/surftemp/.

rising to produce these results. Three, satellite technology has improved significantly in the last ten years. And four, researchers from a company based in Santa Rosa, California published results of a more recent study in the respected journal Science in 2003 that significantly weakens the arguments against global warming that had for a decade relied on the results found by the Huntsville researchers. For the first time in more than a decade, satellite data was analyzed extensively once again. But this time, using more sophisticated satellites measuring temperatures in the upper atmosphere throughout the entire planet found that average temperatures in the upper atmosphere are *nine times higher* than what was claimed by the Huntsville researchers. These results mesh closely with actual worldwide surface temperatures, ocean water temperatures, and polar icecap temperature readings, proving that nearly all climate change computer models being used today are accurate in their assumptions of what the temperature change should be in the upper atmosphere.

One can now only question what is *causing* global warming – not whether it is actually occurring. Therein lays the only remaining debate. However, the science of measuring human impact to global warming has become markedly more powerful in recent years.

Examples of methods used to accurately measure historical climate trends – and their causes – are:

Pollen grains: Deep core drilling of mud samples from ancient lakes allows us a near-perfect window into the past climate. The rich brown color of the earth taken from these core samples is full of organic matter, especially pollen. Under powerful microscopes, these pollen grains can be studied to determine how the species of vegetation in the area has changed during climatic variations. Each meter of mud contains 2,300 years of pollen grains from trees, grasses, and flowering plants. Some lakes allow continuous sampling as far back as 42,000 years (the last ice age occurred 21,000 years ago). Some lakes since buried by glaciers provide records dating back over 100,000 years. "Here's what the forest looked like 21,000 years ago at the height of the last ice age," says Cathy Whitlock, a fossil pollen expert and paleoclimatologist at the University of Oregon. "And, oh man, was it a different world."

Ice core samples: Tropical mountains, such as Kilimanjaro and those in Peru have glaciers that are thousands of years old. Although these glaciers are dwindling due to melting, the stable center of these glaciers have not yet melted and hence provide an accurate historical record of atmospheric gases contained in the frozen ice and moraines. "We want to understand how the climate worked before and after people appeared on this planet. That's the only way we'll figure out what impact people have on climate, how much we're responsible for the way it's changing now" says Ohio State University glaciologist Lonnie Thompson. Antarctic ice core samples date back over 400,000 years and suggest a strong link between ice ages and astronomical rhythms, as well as drops in the CO_2 levels in the atmosphere during ice ages. Global warming gases can be accurately measured over hundreds of thousands of years, including over 10,000-year ice age cycles, the Earth's 23,000-year and 41,000-year global axis tilt cycles, and the Earth's 100,000-year sun-orbital cycle. This allows researchers to study the correlation between climate change and global warming with a remarkable degree of accuracy. The results: never before in the history of mankind have CO_2 levels been as high as they are today – ever. We know how much CO_2 we produce from power plants and vehicles, since it is monitored and regularly calculated. We also know that planet Earth has never had to endure so much CO_2 or other global warming gases. This strongly suggests that our massive production of global warming gases is significantly impacting the climate.

Stalagmites from caverns: Caves in the Midwest can provide a trove of past climate data. By measuring isotopes of uranium, as well as mud embedded in the slow-growing stalagmites, scientists can look back 10,000 years at climate change events, rainfall patterns, and flooding in the caves.

Pack Rat nests: Pack rats have been hoarding seeds, twigs, and leaves in the arid Southwestern United States for over 30,000 years. Preserved in crystallized urine, these samples provide a record of climate change in that region.

Corals: Corals produce annual rings like trees do. The density of a ring depends on sea surface temperature, which has risen dramatically near Bermuda over the last 50 years according to Woods Hole researcher Anne Cohen. "Those temperatures were pretty steady from 1850 to 1950. Then boom! They just shot up. You can see global warming right there," she says.

Tree rings: Tree rings like those in a hemlock buried for a millennium and uncovered in Alaska by the retreating Columbia Glacier provide annual temperature readings from A.D. 585 to the present. "Living trees seem to be experiencing stresses they haven't seen in the past thousand years," says geologist Greg Wiles. Some of the sequoia trees in northern California live over 2,000 years, providing an accurate looking glass into the past climate of that region.

Other methods: Other methods include dust and sand dunes, microscopic shells of organisms buried in deep-ocean sediments which are studied with deep-core drilling techniques, archaeological inscriptions, ship captain's logs, and ancient vintner and gardening records. Black bands were found in a Kilimanjaro ice core: "that's dust," says Lonnie Thomson, dating back "to 4,200 years ago when there was a terrible drought in North and East Africa. The air was full of sand, dirt, and dust…which mixed with the snow as it fell on Kilimanjaro." These dust records, when compared to hieroglyphic inscriptions from the period, describe how the Egyptians suffered a 50-year drought along the Nile; many died from famine. The Old Kingdom, described in religious texts, ended during this time period. Thompson and some European history scholars believe the severe drought spread north into the eastern Mediterranean region and contributed significantly to the collapse of both the Old Kingdom and the Akkadian empire in Mesopotamia. Although the dust was clearly caused by the drought and its related natural sources, "it shows what climate change can do [to a civilization]," says Thompson.

These and other records provide irrefutable proof that global warming gases released by mankind are the primary cause of global warming. And we all know that the burning of fossil fuels is the primary source of these gases.

GLOBAL WARMING'S IMPACT ON MANKIND

There is no doubt that global warming is wreaking havoc in visible, as well as more subtle ways that we may not have discovered yet, and that this damage is occurring throughout the planet. Such trends force our nation – and nearly all others – to pay an ever increasing percentage of GDP towards the costs of global warming. These costs are both direct, which are easily quantified, and indirect. The indirect costs to our economy of global warming, the true nature of which are only now being discovered and are more difficult to measure, remain elusive. But research suggests that the true cost to our economy of global warming could become staggering in less than a decade.

Higher carbon dioxide levels were recently detected over Antarctica. The average airborne CO_2 levels over that continent have risen by 2.6 percent in just six years, according to measurements taken by Japan's National Institute of Polar Research with weather balloons hovering 9 to 19 miles above the surface. This is the first greenhouse gas increase above the South Pole that has ever been detected, with carbon dioxide up 9.4 parts per million (ppm) from 1998.[62]

---

[62] USA Today, September 29, 2004, p. 6D.

The Kyoto treaty, with Russia signing in 2004, will go into effect officially in 2005. This comes as anxiety among America's business community increases relative to the high cost and low availability of energy. DuPont Company, as an example, has curbed carbon dioxide emissions by 67 percent since the Kyoto treaty came along – on a voluntary basis. The company says these reductions have not cost more, but have actually made their plants more efficient and "positions the company for the marketplace of 20 to 50 years from now."[63] This trend creates opportunity for those companies offering energy efficiency products and services, as well as clean power technologies. Ultimately, this will not harm the American economy, but instead will save it from much greater pain later.

Temperatures have been tracked for hundreds of years. The hottest years on record (as of mid 2004) are:

1. 1998
2. 2002
3. 2003
4. 2001
5. 1997

The 1990s were the hottest decade of the millennium. In 2003, the third hottest year on record, Europe was hit especially hard. Europe saw its hottest summer ever cause over 15,000 more heat related deaths in France during August 2003 than what is expected during a typical August,[64] over 4,000 more died in Italy in that country's 21 largest cities, as well as thousands more in Portugal, Spain and other southern European countries. Record heat caused electricity demand to surge due to higher air conditioner use, and there was significant damage caused by more frequent forest fires and property fires. Switzerland saw its hottest June in 250 years during the summer of 2003.[65] Higher temperatures and more frequent heat waves could double the number of heat related deaths in Atlanta, Georgia by 2050. Similar trends could worsen if temperatures continue to climb.

WARMING TRENDS

The ski, snowboard, and other winter sports industries could be hit hard by continued warming. Already, many resorts have shortened seasons and

---

[63] Los Angeles Times "Industry Energized by Kyoto Pact" by James Flanigan, 10/10/04.
[64] Estimate from France's National Institute of Health and Medical Research.
[65] Science News, "Dead Heat" July 3, 2004, p. 10.

increased dependence on manmade snow. In southern Ontario the length of the ski season could be cut in half by 2080.

Rainfall could decline by 10 percent in drought-plagued Ethiopia. In Africa, widespread poverty and dependence on subsistence farming make that continent the most vulnerable to climate change.

Higher temperatures could dramatically increase demand for air conditioning in Florida and other warm regions of the world, increasing demand for electricity produced today primarily by burning fossil fuels.

Intense hurricanes may occur more frequently as warmer waters in the Atlantic make it easier for them to form and make them more powerful. This prediction, written in August 2004 for the story in National Geographic, has proven to be eerily accurate since the 2004 hurricane season is by far the worst in history, costing Florida well over $20 Billion of insured damage, and up to an estimated $40 Billion in overall damage. Islands in the Caribbean, where many American companies that have invested billions in real estate, saw massive property damage – and in some cases complete destruction.

The American Midwest would become warmer and drier, lowering crop yields. Every one degree increase in temperature is estimated to lower corn yields by 10 percent.[66]

China, Southeast Asia, and the western United States may see more precipitation but less snowfall, jeopardizing the drinking water of millions in cities like Los Angeles and San Francisco.

Sea levels are projected to continue rising, causing extensive property damage in cities like New York, New Orleans, Miami and many others around the world.

Malaria and other ailments previously relegated only to the tropics might be seen as far north as Vermont.

Fish populations, even in areas where fishing limits are strictly enforced, have seen a dramatic decline in recent years. This trend is expected to worsen as water temperatures continue to rise. Over-fishing and climate change, along with rising demand for seafood, are resulting in smaller average fish weights and smaller numbers caught.

---

[66] According to ecologist Christopher B. Field of the Carnegie Institution.

Researchers announced in 2004 that the tropical Atlantic Ocean is much saltier than it was 50 years ago. For years, scientists assumed that global warming would speed evaporation in parts of the world's oceans, but they had no direct way of measuring the change. In the prestigious journal Nature, the Woods Hole Oceanographic Institution in Massachusetts studied salinity over several decades. They found a 10 percent rise in tropical evaporation over just the last 15 years. They also found a corresponding freshening of water in the far northern and southern regions of the Atlantic, adding credence to the theory that fresh water is being added to those regions from polar ice water runoff as temperatures rise. As the oceans become less salty, the ability of Earth's oceans to absorb excess carbon dioxide decreases. In 2001, the ALOHA weather station 100 kilometers north of Oahu showed that the ocean absorbed only about 15 percent of the CO_2 that it did 12 years earlier in 1989 as salinity increased by one percent.[67] Since the ocean has been absorbing about one-half of the world's carbon dioxide, this steadily diminishing ability to literally soak up fossil fuel emissions spells trouble for the next few decades. It could actually lead to a slight increase – or geometric acceleration – of the warming trend as less CO_2 is sequestered by the ocean every year.

U.S. water supplies, already severely contaminated in places like Atlanta and Washington D.C., are becoming more and more strained. Many studies are showing that there will be a need to spend billions of dollars in the coming years in America alone just to maintain current standards. But if warming trends continue, water coming from rivers, streams, and reservoirs may decline. This could cause the pollution levels in water to become more potent. As toxins caused by electricity generation rise while water levels drop, pollution could get much worse if rising power plant emissions are not capped.

According to a study released in November 2004,[68] arctic sea ice has decreased 8 percent in just the last 30 years, resulting in the loss of 386,100 square miles of ice. This is an area, as large as Texas and Arizona combined, that no longer has ice cover.

Ice on Lake Mendota near Madison, Wisconsin now covers the lake 40 fewer days than it did 150 years ago. University of Wisconsin researcher John Magnuson says, "Wisconsin is losing winter as we knew it."

---

[67] Science News magazine, August 16, 2003, p. 101.
[68] ACIA report, Cambridge University Press, released November 9, 2004.

SPECIES IMPACT

Global warming and climate change affect our friends in the animal kingdom just as severely. More than 11,000 known species of animals and plants are currently at risk of extinction. The destruction of the rain forests continues to occur even now at an alarming rate of some 30 million acres per year.[69] In addition to eliminating one of the Earth's most powerful mechanisms to absorb carbon dioxide, rain forests contain a plethora of rare and exotic species that exist nowhere else on Earth. Some of those plants, insects, and animals could hold the clues to miraculous cures to disease and aging. Global warming is the worst contributor to worldwide species decline, even more so than the gradual destruction of the vast rain forests. A few disturbing examples are:

The Fleischmann's glass frog in Costa Rica: Numbers plunged inexplicably in the late 1980s. Half of that region's frog species have vanished or declined. The world's amphibians are seeing a steady decline in population.

The ice-dependent Adelie penguins on the Antarctic Peninsula: Population has dwindled by 66 percent in 14 years due to a rise in average winter temperatures of 9ºF over the last few decades. Sea ice has retreated by 20 percent since the mid 1970s, which deprives the penguins of their primary feeding platform. They are being replaced by warmer climate Gentoo penguins, which have increased 55-fold since the early 1990s. "The Adelies are the canaries in the coal mine of climate change in the Antarctic" says ecologist Bill Frasier. "I've watched this area over 30 years. I'm in awe that it has taken such a short time [14 years] to happen."

The Edith's checkerspot butterfly: Once ranging over 300 miles in southern California and northern Mexico, the Edith's checkerspot butterfly has become extinct in 80 percent of its historical range due to rising temperatures.

Coral Reefs: Heated waters trigger coral to shed the algae that nourish it, a bleaching event that leaves coral white. Some reefs recover, but many die and become briny boneyards. Dead coral offers a poor habitat for passing fish and other marine life. Australia's Great Barrier Reef is the world's largest at 1,240 miles long and is home to some 400 species of coral and 1,500 species of fish – the ocean's equivalent to a tropical rain forest. In 1998, the hottest year in history,

---

[69] U.S. News and World Report Special Edition "The Future of the Earth: A Planet Challenged from the Arctic to the Amazon" July 2004, p.5.

the world's coral suffered its worst year on record leaving 16 percent of the world's coral bleached or dead.

Vegetation: Plant life in the Rocky Mountains, Adirondacks, The Great Smoky Mountains, Appalachian Mountains, the Andes, the Alps, the mountains of southern Switzerland, and other areas is moving upslope as temperatures continue to rise. Exotic plant species are invading the areas where these cool-air-loving plants once thrived.

Shellfish: Can be damaged by carbon dioxide. Nearly half of the CO2 that mankind has spewed into the air in the past two centuries has ended up in the oceans. This could damage shellfish ability to make their shells. In fact, CO2 in the ocean can cause the shells to dissolve completely, putting at risk such creatures to make their shells. CO_2 in the ocean can even cause the shells to dissolve completely, putting at risk such creatures as corals, snails, plankton, crab, lobster and other shellfish.

Polar bears: Those living in the North Pole may become completely extinct by the end of this century since the ice cover over water where seals live and play may completely disappear.[70]

Mass Extinction: In the first study of its kind, research published in the journal *Nature* showed that global warming at its current rate will drive 15 to 37 percent of all species toward extinction by 2050, only a few decades away.[71] "The midrange estimate is 24 percent of global species committed to extinction by 2050. We're not talking about the occasional extinction. We're talking about 1.25 million species. It's a massive number," says ecologist Chris Thomas of the University of Leeds in Britain. The 19-member international study marks the first time scientists have performed a global analysis showing concrete estimates of warming trends' impact to species.

Countless other stories of species decline, extinction, and impact due to climate change are appearing all over the world. It is unknown exactly how many species have already been lost forever to extinction in the last 15 years, or how many more will vanish each year from now on. What ecologists do know, however, is that unless we take immediate and massive action to curb global warming, things will only get worse.

Portions of the previous sections were reprinted with permission from National Geographic.

---

[70] ACIA report, Cambridge University Press, released November 9, 2004.
[71] Washington Post "Warming Imperils Species, experts say" by Guy Gugliotta, 1/8/04.

A CHILLING REMINDER[72]

In the middle of the south Pacific, over 2,000 miles from Chile and over 1,250 miles from its nearest neighbor, lies one of the most isolated places on Earth: Easter Island. Best known for its colossal stone statues, this 150-square-mile chunk of volcanic rock teaches us a valuable lesson about the use of natural resources.

The Rapa Nui tribe cut down every tree on the island to construct and move these massive stone heads in the 16th century, leaving behind a barren landscape devoid of trees. As the palm trees disappeared, so did the birds and animals that lived there. And so did their supply of firewood, shelter construction materials, and wood for making fishing boats. Competing for dwindling resources, the tribes began warring with each other. By the end of the 17th century, the Rapa Nui had destroyed not only their ecosystem, but also each other.

Historians tell us that when the island's population reached its peak of 10,000, the number of people greatly exceeded the resources available to sustain them. Surely they realized this, but yet they continued to ravage their land until eventually it wiped them out.

Earth is an island as well, in the cosmos of space. As our global population swells from 6 billion to a projected 9 billion by 2050, and as the impacts of global warming continue to reduce our ability to support ourselves, will we learn from the Rapa Nui of Easter Island? Or will we repeat their mistakes by ignoring the

---

[72] U.S. News and World Report Special Edition "The Future of the Earth: A Planet Challenged from the Arctic to the Amazon" July 2004, pp.4-5.

warning signs all around us, only to leave our children and grandchildren a broken planet unable to sustain life as we know it? I pray that my descendents will not be required to 'duke it out' with their neighboring nations to control the last remaining resources.

CONCLUSION

We do not live *on* the Earth, but *in* it. The vast atmosphere swirls around us, affecting everything. Man is increasingly affecting our own climate, and Mother Nature is losing her ability to buffer our short-sightedness. No matter how many vitamins you take and no matter how many health and wellness measures you employ, you are still breathing polluted air, drinking polluted water (most restaurants do not filter their water, plus this unfiltered tap water is used in your coffee, tea, and other beverages), and eating toxins in your food (most of the water used to prepare your food is unfiltered, and even organically grown foods contain the poisons from burning fossil fuels inherent in the soil and water used to grow them).

The only way to escape pollution from the burning of fossil fuels is to move to a remote mountain hideaway, grow all your own food locally using organic methods, filter all your water, and stop breathing. Even in places like Hawaii and the mountaintops of Montana the air is polluted – many of these toxins blow along in the air for thousands of miles, floating to the most remote areas of the world. Pollution from power plants in China floats all the way across the Pacific Ocean to the western United States and is being measured at protected redwood and sequoia forests in California. <u>The Economist</u> says "Pollution in China is reaching scandalous proportions. The World Bank recons that China is home to 16 of the world's 20 most polluted cities, and calculates that pollution and environmental degradation together cost China as much as $170 Billion annually."[73] The U.S economy is far larger than China's, so it may be fair to say the cost to our nation of pollution could equal or exceed $170 Billion. This subject of economic cost is addressed in detail in Chapter 7. There are other hidden costs as well: estimates suggest that 300,000 people die each year in China prematurely from respiratory diseases.[74] This is due primarily to coal-fired generation: 70 percent of China's rapidly growing energy needs are met with coal, and a quarter of the country endures acid rain. Yellow dust clouds were so extensive in 2001 they raised complaints in South Korea and Japan and traveled

---

[73] <u>The Economist</u> magazine, August 21-27, 2004, "China's Growing Pains" pp. 11-12.
[74] The Economist magazine, "A Great Wall of Waste" p. 56.

as far as America. This is truly a global problem – one from which there is only one escape: stop burning fossil fuels.

Our energy costs in America are artificially low because we do not accurately reflect the health and environmental costs of burning fossil fuels. These pollutants are slowly poisoning all of us and warming up the Earth. As we have seen, this damage to our health is diminishing our quality of life and actually shortening our lives. It is time to seriously reflect on these problems, and debate what should be done about it. Fortunately, a cost-effective solution exists today, as you will soon learn.

As the global warming debate rages on, its character has changed of late. No longer do opponents debate whether or not global warming is occurring. The signs are simply far too ubiquitous to ignore.

The debate about whether human beings have an impact on global warming has also lost its appeal, since those signs are all too obvious as well. The debate now centers not on *whether* mankind has an impact on global warming, but on *how much* of an impact we have. No one knows for certain. But in 30 years we will know a lot more.

Personally, I do not care to waste any more time debating the issue. I have seen enough. It is time to stop talking about it and do something to make a difference.

Chapter 4:

WINDS OF CHANGE

The 3 states of Kansas, Texas and North Dakota have enough wind potential to power the entire country.[75]

When I read that fact in 1998, it blew me away. Just 3 states could power the whole nation? I thought at the time, 'if that was true, why aren't we building wind farms like crazy?' So I began researching the wind energy industry, and quickly discovered that not only was it true, it was an understatement. In fact, Stanford University proved in a 2003 study[76] that the winds of the Midwest could power the entire U.S. electric grid 8 times over!

This chapter will hopefully dispel some of the myths about wind power and educate you about the basics of wind energy generation. Along the way, I hope to convert you to a wind energy supporter and enthusiast. If nothing else, perhaps you will vote to support political candidates that promote more development of clean, renewable, American-made wind power. Or if you live in another country, I encourage you to support wind energy legislation and incentives with your vote and with your pocket book.

As of January 2004, worldwide installed wind energy powers the equivalent of 9 million average American homes (equivalent to roughly 30 million people, or more than New York City, Los Angeles and Chicago combined). Since European homes are more efficient than U.S. homes, the world's wind energy production could power 19 million European homes. About $9 Billion was invested in new wind energy equipment in 2003 alone, up from $7 Billion in 2002.[77]

---

[75] Department of Energy study (Pacific Northwest Laboratory) (1991): www.pnl.gov.
[76] Stanford University study published in the Journal of Geophysical Research (May 21, 2003): http://news-service.stanford.edu/news/2003/may21/wind-521.html.
[77] American Wind Energy Association "Global Wind Energy Market Report"

WIND HISTORY

Wind energy has been used for thousands of years, dating back even to ancient times. Boats have been powered by the wind catching their billowing sails for millennia, and the Dutch built enormous wind mills to grind grain and pump water.

Image of an ancient wind-powered grain mill.
Note the size of the doors to the structure and the vertical axis of the windmills.
Image Courtesy NREL

WIND POTENTIAL

Is the wind powerful? The F5 (as rated by the Fujita scale) tornado that swept through Oklahoma City in May of 1999 set record wind speeds of 318 miles per hour.[78] F4 tornadoes are described by the National Weather Service as "devastating" and a F5 tornado, called "incredible," is strong enough to rip the tarmac off roadways and lift strong frame houses off their foundations and carry those homes a considerable distance.[79] F5 tornadoes hit the United States only about once per year. Tornadoes, which are usually far more powerful than hurricanes or typhoons, cause catastrophic damage wherever they near the ground. Mother Nature's winds have enormous power: the energy contained in the wind on the ground of the May 1999 Oklahoma City F5 tornado was equivalent to the shock wave of a nuclear bomb (as reported by The Science Channel in a special aired in September 2004). Yes, the wind is powerful.

---

[78] CNN: http://www.cnn.com/WEATHER/9905/07/okla.tornado.02/.

[79] National Weather Service website: www.nws.gov.

The map below shows the windiest areas of the United States.[80] As you can see, the Midwest has the greatest potential for wind power. The high plains area (western Kansas, eastern Colorado, north to the Dakotas and south to the Texas panhandle) are often called 'the Saudi Arabia of wind.' The Midwest has enough wind potential to power the entire U.S. electric grid eight (8) times over. Put another way, the Midwest could power the entire transmission grid and have enough wind resource leftover to produce enough hydrogen (using wind and water) to power every car and truck in America.

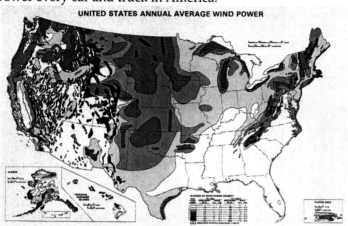

To produce 15 percent of America's electricity (double what hydroelectric dams provide today), only 0.6 percent of the land in the lower 48 states would have to be developed with wind power plants. To meet 100 percent of America's yearly electricity needs, or 4,000 terawatt hours, using wind (the United States produced 3,841 terawatt hours of electricity in 2002, which was a 2 percent increase over 2001) wind turbines would need to be spread out over no more than about 15 percent of each of the windiest states. These wind farms would be located primarily in rural areas where the positive economic impact is needed (and wanted) the most. Keep in mind that within these areas, only about 2 percent of the land would be taken out of agricultural production, so farming and ranching could continue in large part unaffected by wind farms. Additionally, wind resource is now known to exist outside of the Midwest in the ocean offshore the southeastern United States. This previously uncharted area has now been documented and shows developable wind potential; suggesting that Florida, the Gulf Coast states, Georgia and the Carolinas could be powered in large part by the wind as well.

---

[80] More detailed wind maps are available, including DOE Wind Powering America.

Wind power installations in the U.S. are geographically diverse:

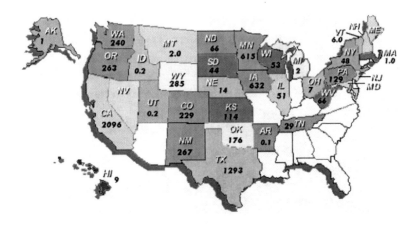

New transmission lines are required to move large quantities of power from rural areas to population centers. Some of these power lines can utilize technology that emits lower EMF (electromagnetic fields, known to be linked to health problems) and can even include fiber optics lines for sophisticated monitoring to enhance reliability of the grid (see Chapter 6 to learn how the Freedom Plan can convert the entire nation to clean, renewable power).

Wind energy is the fastest growing form of power in the world.[81]

Global additions and cumulative wind power capacity (MW) (3)

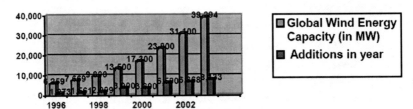

Image Courtesy AWEA

And, wind energy is the fastest growing form of power in the U.S., growing at an average annual clip of 28 percent per year.[82]

As you can see, in the U.S., wind power growth has been choppy. This is due to the inconsistent federal policy driver (the Production Tax Credit, explained later in this chapter). Wind energy growth in Europe has been consistent and strong.

---

[81] American Wind Energy Association (AWEA).
[82] AWEA.

Germany, the largest economy in Europe, now leads the world in wind power having invested over $24 Billion in new wind facility additions since 1990 (compared to about $8 Billion in the U.S.). Some states in Germany are powered over 30 percent by wind energy, while the entire country (geographically roughly the size of Florida) is powered nearly 10 percent by the wind. There are enough wind turbines in the country of Denmark that at times the wind blows strong enough to power 100 percent of that country (on average over 40 percent of Denmark's power comes from the wind). Spain, Great Britain and other members of the EU are rapidly gaining ground on the U.S. (Spain is expected to surpass the U.S. in total wind power installations in 2004). Denmark has announced plans to power 100 percent of its electric needs with wind year-round. Great Britain announced plans to power 20 percent of the country with renewables (primarily wind) by 2017.

Wind power growth is accelerating rapidly in China, India, and many other countries around the world. Yet the United States, with the most wind potential of any nation on Earth, is lagging far behind most other developed and developing countries in its utilization of clean, renewable wind power. With consistent federal policy and enough public demand for wind power, the USA could once again regain the lead in global wind energy production. But government policy alone will not convert America to clean power – only public demand for immediate and massive action will make it happen.

COST

The cost of generating a kWh (kilowatt hour) of electricity from wind power has dropped by 90 percent since 1980.[83]

Cost of Wind-Generated Electricity 1980 to 2005, Levelized cents/kWh

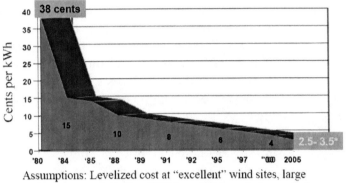

Assumptions: Levelized cost at "excellent" wind sites, large project size, not including PTC (post 1994), costs in nominal cents/kWh.

Image Courtesy NREL

Today, large wind farms (200 MW or larger) built at excellent wind sites (those with mean wind speeds of greater than 9 meters per second or 18 miles per hour) can produce energy at 3 cents per kilowatt hour (kWh) without the federal production tax credit (PTC), and 1.5 cents or less per kWh with the PTC. This price is the energy price delivered at the busbar or substation, and does not include any additional transmission costs. It also assumes an excellent wind resource (capacity factor of 39 percent or better), a low cost of capital to finance the project (blended rate of 10 percent or better), interconnection costs (building the substation to connect to the grid) that do not exceed 10 percent of the total installed cost of the project, and an installed cost of about $950,000 per MW.

For comparison, a 2001 study performed by Xcel Energy (the 5th largest utility in the USA) determined that building a new coal-fired power plant could produce energy at 3-4 cents per kWh, a new natural gas-fired plant could deliver energy at 3.5-5.5 cents per kWh.

A large wind farm built in Oklahoma and owned by Florida Power and Light sells its power to local utilities at a cost of 1.85 cents per kilowatt hour ($18.50/MWh). This sub-two-cent price is now becoming the norm for large wind farms in the Midwest that are utilizing the Federal Production Tax Credit

---

[83] NREL (National Renewable Energy Laboratory).

(without the PTC the cost would be just over 3.5 cents per kWh, still less than building a new coal or gas plant).

That makes wind power the least expensive form of <u>new</u> electricity generation.

But wind energy is more expensive than the energy from existing coal plants, which can deliver energy prices of 1-1.5 cents per kWh. These plants, however, do not have the modern emission control systems required at new plants (they are 'grandfathered' in terms of emissions requirements) and have often paid off the loans used to finance them. And, the true cost of that electricity is not accurately reflected, since the health damage, eco-system clean-up and environmental costs are not currently reflected in the wholesale price of coal-generated power. Therefore, the raw economics of burning coal for electricity at existing power plants is less expensive than wind power – today.

In remote areas, wind energy potential can be even more cost effective than fossil fuels due to the added costs of shipping fuel or building power lines to the area. Researchers estimate that wind can produce more electricity at less cost than diesel generators at remote sites with average wind speeds of 4 meters per second (about 9 miles per hour) or greater.[84] A typical village household in a remote area of the world will use 60 to 300 watts of power at any given time, and use up to 1,200 watt hours (1.2 kWh) per day, or 40 kWh per month. Therefore, a small 3 kilowatt wind turbine costing $15,000 to $25,000 installed could provide enough power for 10-30 village homes with zero pollution, very little maintenance and at least a 20-year life.

Matthew Patsky, portfolio manager for the Winslow Green Growth mutual fund, says "Wind broke to a point of being economical in most parts of the U.S. vis-à-vis the cost of electricity back when oil passed $30 per barrel, and now it's incredibly economical. I don't think there's very many places in the country where there's a cheaper way to produce energy than wind."[85]

The U.S. Energy Department estimates that 600 GW (gigawatts) of new wind capacity (20 percent of U.S. electricity needs, or equivalent to what nuclear power provides now) is cost effective when natural gas is $4 per decatherm (gas has averaged about $5 per decatherm in 2004). This is a conservative estimate since the U.S. Department of Energy uses slightly higher cost figures for wind

---

[84] American Wind Energy Association (AWEA) "Wind Energy Application Guide" p. 7.
[85] "Not Many Alternatives for Energy Investors" by Meg Richards, AP, NY, 10/2/2004.

energy than what is actually the case. For example, the DOE says wind energy is around four cents per kWh today.

States developing renewable energy have found it costs less than expected and typically saves money, according to PRNewswire.[86] Some utilities resist renewable portfolio standards (RPS) that require them to purchase clean power. Amendment 37, which passed in November 2004 in Colorado, was the first time in the country that a state's citizens bypassed their own state legislature (seen as ineffective in this area) to require power companies to switch to more sources of clean power. But some utilities objected vigorously, claiming it would raise their costs.

"This assertion is totally unsupported by actual experience," said Jon Wellinghoff, author of Nevada legislation requiring utilities get 15 percent of their energy from renewables by 2013. Before the Nevada legislature unanimously adopted its RPS, said Wellinghoff, utilities claimed their costs would increase $300 million. They have since testified to the Nevada Public Utilities Commission that they expect their first renewable contracts to *save them $15 million* over 20 years.

This mimics what happened with the RPS that passed in Texas in 1999. Utilities and industrial customers initially resisted the bill, but then realized it saved money as large wind farms created economies of scale in development work and equipment costs. Texas quickly exceeded its clean power targets and is now one of the leading wind power states in the country. Wind power in Texas now costs less than gas-fired power, said Mike Sloan, who also helped craft the Texas renewable standard. In California, where a RPS was enacted in 2002, early clean power project proposals are coming in at a low enough cost that utilities will not need to tap public funds that are available under the law, said Ryan Wiser of the Lawrence Berkeley National Laboratory.

There are several myths about wind power that need to be dispelled. Some opponents to wind have spent a great deal of time and energy dispersing misinformation about modern wind energy. Therefore, the following section is intended to dismiss those myths and arm proponents of wind energy with some of the data and facts they need to make their case.

---

[86] "Renewable Energy Laws Saving States Money" PRNewswire, Seattle 10/26/2004.

10 Wind Power Myths

People mistakenly think that Wind Power:

1. Is Unreliable
2. Kill birds
3. Is Noisy
4. Is Unsightly
5. Is Dangerous
6. Is expensive compared to fossil fuels
7. Does not produce energy when needed (during peak hours)
8. Is challenging to connect to the power grid
9. Reduces valuable reactive power (VARs)
10. Requires backup power generators running at all times

A brief explanation follows explaining how and why each of these wind power myths is not true. Some of the information may be too technical for some readers, so feel free to skip the more technical sections to make your reading experience more enjoyable.

1.   Unreliable?   Although common sense may tell us that wind power is intermittent, and therefore unreliable, few people realize that wind patterns are highly predictable.   In fact, wind forecasting software has gotten so good, modern wind farm operators can determine exact wind speed (and therefore wind energy output) to within plus or minus 5 percent up to 24 hours in advance.   A wind farm operator can predict the output of his/her wind farm better than the local utility can predict what their demand load will be.   And, sophisticated modeling at a potential wind energy site can provide a surprisingly accurate projection of average hourly output over next 30 years.   This forecast allows the local grid operators to reliably predict what their energy needs will be when the wind dies down to schedule other sources of energy production.   In most good wind areas a typical wind farm will produce power at least 90 percent of time (in other words the wind is either not blowing hard enough or blowing too hard to be within the performance range of the wind turbine only 10 percent of the time).   Wind turbines built after 1998 generally begin producing energy with wind speeds as low as 2.5 meters per second (about 5 miles per hour) and can operate in winds up to 25-30 meters per second (about 50-60 MPH).   This wide range makes them increasingly more efficient than older wind generating units.   And, wind farms are available (not down for maintenance) greater than 99 percent of the time (compared to 72 percent availability for coal-fired power plants[87]).

Example of a real-time wind farm forecast:

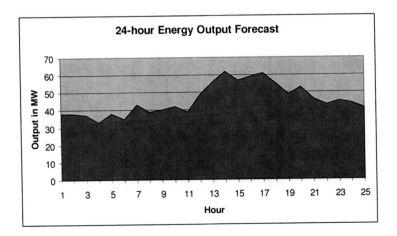

[87] "Benchmarking Air Emissions" published April 2004 by NRDC, CERES, & PSIG, Inc.

80

Actual Wind Forecast Data:

120 **MW Nameplate Capacity Wind Farm**
REAL-TIME Forecast (updated hourly)

HOUR	Predicted Output (MW)	Capacity Factor	Margin of Error
0	38	38%	+/- 0%
1	38	38%	+/- 1%
2	37	37%	+/- 2%
3	33	33%	+/- 2%
4	38	38%	+/- 2%
5	35	35%	+/- 3%
6	43	43%	+/- 3%
7	39	39%	+/- 3%
8	40	40%	+/- 3%
9	42	42%	+/- 3%
10	39	39%	+/- 3%
11	49	49%	+/- 3%
12	56	56%	+/- 3%
13	62	62%	+/- 5%
14	57	57%	+/- 5%
15	59	59%	+/- 5%
16	61	61%	+/- 5%
17	55	55%	+/- 5%
18	49	49%	+/- 5%
19	53	53%	+/- 8%
20	46	46%	+/- 8%
21	43	43%	+/- 8%
22	45	45%	+/- 8%
23	44	44%	+/- 8%
24	41	41%	+/- 8%

2. Kill birds? Some environmentalists are concerned that wind farms will kill birds and/or disrupt the migratory patterns of birds since modern wind turbines stand around 80 meters tall (roughly 300 feet) with the tips of blades going 120 meters in the air (about 400 feet) or higher. This is a valid concern, but one that has proven time and time again to not be an issue. The one place in the world where wind farms have had a negative impact on avian species is in Altamont Pass, California near San Francisco. This wind farm was originally built in the early 1980s when zoning and permitting regulations did not require an environmental assessment or impact study. As a result, the developers realized much to their dismay that the wind plant shared the site with a nesting habitat for raptors (falcons, Golden Eagles, hawks, etc.). As a result, this is the one place in the world where bird strikes from wind turbines have been a problem. However, wind turbines have changed dramatically since then. Today's wind turbines are 10 times larger with rotors that spin much more slowly, making it easy for birds to avoid the slower moving blades. The blades are not made of sharp metal, but of smooth composite materials. Modern towers are sleek and tubular, leaving nowhere to build a nest, unlike the metal lattices towers used in the 1980s that allowed numerous nesting opportunities. Fortunately, the older wind turbines at Altamont Pass site are now being replaced with modern units, which will significantly reduce the negative impact on raptors and other birds there. Nowhere else in the world have wind turbines been shown to kill birds in any greater number than any other standard structure. And, because of the Altamont Pass mistake, significant study has been done to determine the impact of wind farms on avian species. The results worldwide conclusively prove that wind farms, sited away from primary migratory paths and nesting grounds, have either inconsequential or zero impact on avian species.

Bird deaths by cause:

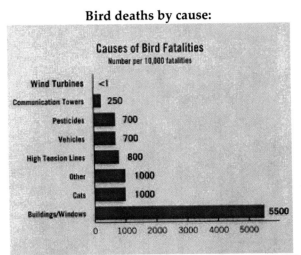

Source: Erickson, et.al, 2002. Summary of Anthropogenic Causes of Bird Mortality

3. Noisy? Modern wind turbines (unlike smaller, older models) operate today at approximately 43 decibels. This is equivalent to the noise of a refrigerator in the kitchen while you are sitting in the next room. And, you have to be very close to the base of the tower (within 100 meters) to even hear the slow "whoosh" sound as each blade passes the tower. Generally, the wind noise blowing in your ear is even louder than the "whoosh" so you rarely hear anything at all. Finally, in nearly every case wind turbines are not built within 150-200 meters of any structures (houses, buildings, etc.).

The average modern wind turbine is over 30% taller than the Statue of Liberty, as shown below.

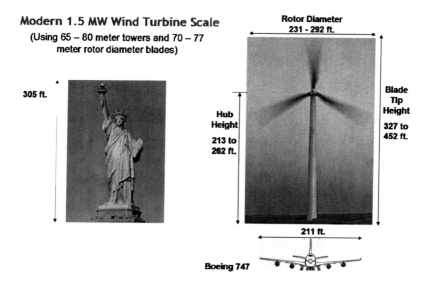

Modern 1.5 MW Wind Turbine Scale
(Using 65 – 80 meter towers and 70 – 77 meter rotor diameter blades)

305 ft.

Rotor Diameter
231 - 292 ft.

Hub Height
213 to 262 ft.

Blade Tip Height
327 to 452 ft.

211 ft.

Boeing 747

4. Unsightly? Beauty is in the eye of the beholder, so if one does not like the way wind turbines look, there is really no argument to be made. Personally, I think wind farms look like graceful, dancing ballerinas. Many people find wind turbines to be mesmerizing yet a small minority (about 2-5 percent) think they are unsightly. There is speculation that some of these opponents to wind farms object to idea for other reasons, such as: resisting change of any kind, being actively involved in the fossil fuel industry, fear of wind turbines scaring away game near their favorite hunting ground, or being afraid of 'ruining' the landscape of certain aesthetically pleasing areas. In my opinion, 10 percent of the population is simply nuts, so no matter what you do there will be objections. Probably the most well-known fight over a wind farm is the Cape Wind project, off the coast of Cape Cod Massachusetts. The opposition efforts there were led primarily by a group of well-to-do homeowners who didn't want to have to see off-shore wind turbines 10-15 miles out to sea (incidentally, from that distance they would be so small that they would be nearly indiscernible). Walter Cronkite, an outspoken opponent of the project, reversed his position in 2003 and began support the project after taking the time to tour the sea where the turbines were to be built and learning the facts about the benefits of wind power. This exciting and needed off-shore wind farm appears to be moving forward now, and will prevent the need for new transmission lines or a new power plant in the Cape Cod area. Polling in Europe, where far more wind farms have been built than in the U.S., shows that support for wind energy tends to strengthen after a wind plant has been installed and operating for some time.[88]

Middelgrunden, Denmark

5. Dangerous? Opponents of wind projects are often vocal and sometimes well-organized, but rarely knowledgeable. Because modern wind energy is relatively unknown, those who oppose it have found that scare tactics and rumors can be effective. At times, ridiculous claims are made that the public may or may not believe, such as:

Ice throws: opponents claim that when the weather is freezing, ice will build up on the blades and the icicles that form will be thrown for miles, possibly going right through your living room window. This is definitely a mere scare tactic, since what little ice that does form has never been 'thrown' outside the fall zone of the tower (within about a 100 meter radius). Thousands of wind turbines exist across the globe, many in cold northern latitudes. It has been proven that the centrifugal force of the turning rotor is the strongest at the bottom of the turn (due to gravity) so any ice that does build up simply falls off the blade straight down to the ground at the base of the turbine. And, many manufacturers now offer black blades (instead of the standard white) and/or heating systems in the blades to minimize icing.

---

[88] AWEA "Wind Power Outlook 2004" p. 5.

Blade throws: back in the late 1970s and early 1980s wind turbines were small, poorly constructed, and had high spinning (high RPM) rotors. At that time, there were occasions when a blade would break off from a little unit and fly through the air, wobbling and flipping, for up to 60-100 meters. With modern wind turbines, this has never happened. Better design and engineering, much, much larger blades that are far too heavy to 'fly' through the air, and slower spinning rotors make this impossible.

Create EMF: electromagnetic fields are created by current passing through aluminum or copper wires. However, high voltage transmission lines (operating at 69,000 volts to 345,000 volts) create the most EMF by far. In fact, in some areas if you stand underneath a high voltage power line and hold a fluorescent light bulb over your head, it will light up without being plugged in due to the high EMF bleeding off the line. Because wind turbines generally have an output of 48 to 600 volts, the EMF is comparable to what is created in the average home, and far less than a transmission or distribution line.

6. Too expensive compared to fossil fuels? As shown earlier in this chapter, wind power is the least expensive form of <u>new</u> energy generation:

200 MW+ Energy Type:	Today's Cost (¢ per kWh)	Estimated Cost in 2006
Wind Farm	1.6 to 2.5	1.2
Coal-fired Power Plant	3.0 to 4.0	3.9
Gas-Fired Power Plant	3.5 to 5.0	4.8

But wind energy can be slightly more expensive than <u>existing</u> thermal power plants (coal, gas, diesel, etc.):

200 MW+ Energy Type:	Cost (¢ per kWh)
New Wind Farm (with PTC)	1.6 to 2.5
Existing Coal-fired Power Plant*	1.0 to 2.5
Existing Gas-Fired Power Plant*	2.0 to 4.0
* Includes subsidies offered to coal and gas power plants	

7. Does not produce energy when needed? Wind farms in some areas have a history of producing power during nighttime hours, when the electric demand is lower. As a result, many utility industry professionals have seized on that fact and claim that wind farms produce power when it is not needed, and cannot produce power when it is needed the most. Balancing the energy production going into the grid with the electric demand is a highly sophisticated and challenging process. Inability to store power on a large scale, combined with the difficulty of forecasting what the demand for electricity will be among millions of residential, commercial and industrial customers, exacerbates the complexity of adding an intermittent resource such as wind power to the grid. However, the myth that wind farms produce power when it is not needed must be dispelled. In fact, there are hundreds of potential wind sites around the U.S. that actually produce more power during the day than at night. Topography plays a large role at each site, and some sites even have characteristics that provide the best energy output during the summer months when air conditioner loads create the highest annual peak demand for electricity. The charts below offer an example of a potential site where 1,000 MW of wind energy could be added to the grid at a delivered price (at the busbar of a substation built at the nearest high voltage transmission line) of only 1.4 cents per kWh.

The best hours of energy production at this particular wind farm site are during the **afternoon** hours when peak demand is at its highest:

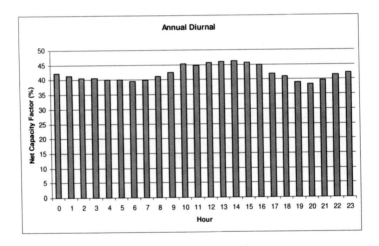

Capacity Factor is the output of a wind turbine relative to its maximum potential. In other words, if a wind turbine is rated at 1,000 kilowatts (or 1 MW), and when placed on a hill somewhere it produces an average of 400 kilowatts over the course of a year, it would have a 40% capacity factor (400 kW is 40% of the maximum nameplate capacity of 1,000 kW). Capacity factor can be calculated for any time period. The current world record for the best average annual capacity factor is 51.7 percent, set in 2002 by a wind farm in New Zealand. Although the average capacity factor in the U.S. has been rising steadily in recent years, it is still low by Midwestern standards at about 35 percent. Many modern wind farms in the Midwest, however, see average annual capacity factors above 45 percent. Older wind farms, like the ones in some parts of California, will see much lower capacity factors since the wind turbines are small and inefficient compared to today's units. Wind energy software has improved substantially in just the last five years, allowing developers to find better sites with improved wind energy output in the form of higher average capacity factors. The table and chart below shows predicted gross capacity factor and energy production based on a single GE 1.5sl wind turbine at a site in the Midwest. As you can see the annual gross capacity factor was predicted to be 47 percent. This translates into a net capacity factor of about 44 percent after incurring array losses (due to the placement of the wind turbines next to each other in a wind farm field, where the downwind turbines will be slightly less productive than the upwind turbines that capture more of the 'virgin' wind flowing through the wind farm) and other losses.

At some sites, **summer** energy output can be very attractive:

Month	80 meters	
	Energy (MWh/yr)	CF
January	7,362	56%
February	7,665	58%
March	8,044	61%
April	7,594	57%
May	7,372	56%
June	9,639	73%
July	7,535	57%
August	6,088	47%
September	7,438	56%
October	6,157	46%
November	6,051	46%
December	5,447	41%
Annual Average	6,160	47%

8. Challenging to connect to the power grid? Because of the intermittent nature of wind, the prevailing theory among utility grid experts, or 'Sparkies' as I like to call them, is that wind energy is challenging, expensive and problematic to connect to the grid. This is not true. A few bad system designs from decades ago do not create a valid reason to claim that the interconnection of wind energy facilities categorically causes problems for the grid system operators. To be sure, there are certain challenges to overcome and guidelines to follow. But today's wind energy systems are generally designed to include a modern substation with all the necessary features demanded by today's Sparkies. These systems include: protection equipment for the grid (if the power goes out and linemen are working on the lines, you do not want electricity from a wind farm to be flowing through those lines), power factor correction (see the next myth), ride through technology for voltage sags, adequate transformers to step up the voltage and 'synch' to the transmission line at the interconnection point, and more. The technical challenges of interconnection of a wind farm to the grid are all well within the capabilities of any reputable electrical engineering design firm, and in many cases since 2001, wind farms actually add <u>more</u> stability to the grid (adding VARs – see below, diversifying generation technologies to protect against fossil fuel shortages and thermal power plant outages due to maintenance, distributed generation puts less stress on the grid, etc.). Therefore, although there are certainly technical challenges, modern-day wind farms are generally helping to improve the reliability and performance of the grid.

9. Reduces valuable reactive power (VARs)? Wind turbines, being 'inductive' machines, are thought to suck reactive power from the grid (also called VARs) which can create system-wide grid reliability issues and cause motors and equipment in peoples' homes and businesses to operate with less efficiency and possibly damage sensitive electrical equipment. Although this may have been true 20 years ago, modern wind turbines no longer consume reactive power. In fact, some manufacturers have patents on their wind turbines that actually add reactive power and leading voltage back into the grid, enhancing the performance and reliability of weak rural, 'radial' or island power lines. And, they offer a low-voltage ride through technology that further improves dependability of the grid:

10. Requires that backup power generators run at all times? This is an easy myth to disprove. At zoning meetings across the country, opponents of wind energy have said that for every wind turbine installed, a matching fossil-fuel generator must be running at all times to 'back up' the wind turbine in case the wind dies down. The need to 'firm the load' as it is called in the utility industry is only necessary when the percentage of generation from an intermittent resource like wind exceeds 20 percent of the total generation in the regional grid. Otherwise, the grid is large enough to balance out the load naturally with current controls, without the need to back-up the wind generation. We shall take this myth apart one element at a time.

First, most wind turbines are rated at 1-2 megawatts (MW), while most gas-fired or steam-fired (coal, diesel, gas, etc.) generators are rated at a far larger 40-500 MW. Smaller 2-6 MW fossil-fuel generators exist, but are usually on site at city municipal electric utility power plants strictly for back up if the larger generators run by larger utilities go down. These smaller, fossil-fuel generators are not very common and are usually older units. Also, they are not cost-effective compared to the large 50 MW+ thermal generators that nearly all utilities prefer. Therefore, although there are thousands of wind farms around the world, there exists not a single utility anywhere in the world which has a wind farm connected to its grid that has a single generator on their system for every single wind turbine. It is irrational logic to claim this myth is true.

Second, most cities have found that running their own small generators costs far more than buying power from the large utility that serves their area with much larger, more efficient, fossil-fired generators. So when these cities keep their small generators on site and maintain them for backup, they rarely turn them on. In fact, they usually turn them on only a few times per year to make sure they are working properly. They <u>never</u> run them at all times for 'backup' because that would be a foolish waste of money and fuel (as well as an added financial cost and administrative headache since they would most certainly be exceeding their EPA emissions standards).

Third, these generators can 'scale up' (turn on after ignition and be fully operational) within seconds. So, if the power goes out from the large local utility (and/or if the grid automatically notifies the computer running the ignition of the small generator that an outage is coming due to maintenance, storms, or any other reason), the small generators can scale up nearly instantaneously and automatically (assuming the town's utility connects to a SCADA (Supervisory Control and Data Acquisition) system to power the town. Therefore, if there were wind turbines nearby in the grid (which almost always employ a sophisticated SCADA system to communicate power output fluctuations, and sometimes real-time output forecasts when needed, to the surrounding grid and generator operators), the generators would operate in the same manner: only igniting and scaling up when necessary, and not running non-stop on idle as some alarmists fighting wind farms would have you believe.

Other myths that common sense tells most people to ignore are: wind farms suck up all the wind and make it hotter in the summer (the winds driving modern wind turbines are 100 meters high, taller than the length of a football field, which have virtually zero impact on winds at lower elevations that can be felt by humans on the ground); wind farms will scare away game, making it difficult to hunt in the area (studies have shown that wildlife readily adapts to these graceful, giant, peaceful structures and that there is virtually no real impact to animals or to nearly all bird species); the generators inside the wind turbines use up more fuel than they save (there is zero fuel use in a wind turbine – there is actually no fuel used by a wind turbine); wind farms cannot make it on their own without tax subsidies (wind farms do not get nearly the tax benefits or direct subsidies that fossil fuel and nuclear generators receive, as explained later in this chapter); and finally it has been said that wind turbines, including their towers and blades, require more energy to manufacture than they generate in their lifetime. Using this last argument would imply that coal-and-gas-fired generators, which require far, far more parts per MW than wind turbines, which require additional parts and infrastructure to deliver a steady stream of fuel to make them work unlike wind turbines, and which have shorter lives than wind turbines, would also require more energy to manufacture than they will produce in their useful life. If this were true, we would have run out of power long ago since the act of making all the parts for new power plants would have used up all our available energy.

Montezuma, Kansas

Image courtesy FPL Energy

WIND DEVELOPMENT PROCESS

To erect wind turbines for commercial electricity generation, a great deal of work goes into what is called the wind development process. This section will outline the basics of wind development activities. For more details, or to learn how to be a wind developer, contact the author or another wind developer. This section is fairly technical, so if you wish to skip forward to the remaining non-technical portion of this chapter, go to the Economic Impact section.

Overview

The typical process used to build a wind farm has five basic steps: 1) site selection, 2) site development, 3) power purchase agreement (PPA) acquisition, 4) financing, and 5) construction. Generally, a wind developer will identify several potential sites for building a wind farm, spend between $500,000 and $1.5 Million per site getting it ready to begin discussions with utilities, and then spend most of their time facilitating the most difficult and frustrating part of the process: getting a PPA from a utility in the region. Once the PPA is obtained, arranging the financing for the project is no easy chore, and then the construction can finally commence.

1. Site selection. Choosing sites to build potential wind farms is much more complicated than licking your thumb and holding it in the air to see if it is windy. There is also much more to it than testing the wind speed and direction, and asking the local landowner if they wouldn't mind selling or leasing their land to you so you can build a wind farm. It should come as no surprise that evaluating potential sites includes modern scientific methods and the evaluation process can cost hundreds of thousands of dollars per site. The factors to consider in selecting a suitable wind site are:

Wind speed: mean annual wind speed, prevailing wind direction, wind shear and turbulence; local airport and meteorological station data can help you ascertain wind resource characteristics at your potential site.

Topography: how the lay of the land will affect the wind coming from the direction of prevailing winds, including trees and structures, keeping an eye out for a 'fetch' – which acts like the underside of an aircraft wing to accelerate the wind over the land – or other attractive topographical features.

Proximity to high voltage transmission lines: keep in mind that building new transmission lines is not only very expensive, but can take years to obtain rights of way permits and approvals.

Availability of excess capacity on those nearby transmission lines: just because a transmission line is nearby does not mean it has any 'room' on it for extra energy.

Environmental issues: make sure your site is not near sensitive environmental areas such as bird habitats or avian nesting grounds – especially raptors or endangered species, native lands which may be considered sacred, archeological sites, scenic byways, etc.

Zoning and permitting obstacles should be acceptable: if other wind developers have been unable to get projects approved, or the history of the local zoning regulatory bodies has been unfavorable to development, it may not be a suitable site; also consider noise and aesthetics issues.

Proximity to customers: the most difficult part, by far, of developing a wind farm is selling the power so make sure there are utilities or other customers nearby that have at least a mild interest in buying the energy that will be generated by your wind farm – and it is wise to check the creditworthiness of such buyers, since a contract with a troubled utility will be nearly impossible to finance.

Ability to lease the wind rights of the land: local landowners may or may not want to see massive wind turbines on their land or their neighbor's land.

FAA permitting issues: tall wind turbines generally cannot be located within 3 miles of any airport, and cannot be within 5-8 miles of the direct flight path of an airport.

Consider state, local, and federal tax incentives for your project: these can have a considerable impact on your ability to get the project financed.

Determine any potential 'fatal flaws' for construction such as the ability to move a 350-ton crane to the area: weak bridges, low overhangs on roads to the site or feasibility of transporting heavy turbines to the site via railroad or durable roads.

2. Site Development. Once a site or sites have been identified upon which to build a wind farm, a wind developer must then begin to develop the site. During this phase a wind developer will often spend between $500,000 and $1.5 Million per site. Examples of development activities are:

Comprehensive wind study: There are really only two things that matter in the wind development business: the quality of your wind data and the PPA. Everything else is just 'fluff' to help support those two primary objectives. The PPA is discussed below, so let us address how critical good wind data is. Since financing a wind farm depends in large part on how much revenue will be generated, the lenders and investors who finance the project will expect to see estimated wind energy output over 20-30 years. The only way to provide that information is to have very good wind data to input into the forecasting software. Thus, if the input of that data is based on incomplete or inaccurate wind resource assessments, the output will be questionable – and hence not attractive to prospective financial partners. To provide 'bankable' energy output forecasts, the wind data must be highly accurate and comprehensive. That requires at least one or two meteorological towers (Met Towers) erected on site, each taking measurements at 10 meters, 30 meters, and at least 50-60 meters at the top. These towers should have sophisticated data loggers taking readings at least every 30-60 seconds, and should be on site for at least 6-12 months. There is no way to circumvent the need for on site Met Towers, since the wind resource can vary widely from site to site, even if the two sites are less than two kilometers apart. Once adequate on-site readings have been collected for a sufficient length of time, that data is entered into sophisticated wind analysis software along with data from nearby airports or weather stations (usually measured at 10 meter height) to correlate the on-site data with a long-term history (20 years or more) of wind speed from the nearby 10 meter Met Towers. Then, additional data can be collected using atmospheric and satellite meteorological models to further enhance the reliability of the wind energy output forecast. Finally, once all the data is crunched by the software, the site is rated based on capacity factor, wind shear, wind turbulence and other factors.

3rd Party Wind Data Certification: Once the wind developer rates their site's wind resource as shown above, an independent 3rd party wind certification company is provided with all the wind data from the on-site Met Towers, nearby 10 meter Met Towers for long-term correlation, and any other data, along with information about the wind measurement collection methods used by the developer (how the on-site towers were set up, where they were placed relative to the prevailing winds and topography, manufacturer and model of the anemometers mounted on the Met Towers, data loggers used, scheduled maintenance and human Met Tower monitoring activities, etc.). This 3rd party wind certification firm then produces all their own reports on this data, providing probability factors that the energy output results will actually be what the wind developer claims they will be. This certification process is performed primarily for the benefit of the lender and any other parties who will be taking out the risk of financing the project. Because the loans are generally non-recourse (the lender cannot penalize the equity owner of the wind project if the wind farm does not produce enough revenue to meet the loan payments), the lenders are keen on the outcome of these probability factors provided by 3rd party certification. It cannot be emphasized enough how important the raw data of the wind resource assessment is for the successful financing of the project.

Transmission Studies: To determine what type of substation needs to be built and other interconnection costs, as well as ascertain whether there are any fatal flaws in connecting the wind farm to the grid in that area, a series of transmission, system impact and interconnection studies must be performed. These are generally performed by either the local utility that owns the transmission line the wind developer wishes to connect to, or by the RTO (Regional Transmission Operator) under regulation by FERC (Federal Energy Regulatory Commission), can take 6-24 months to perform, and often cost $50,000 to $100,000 per study. This study also determines the impact to the grid which may lead to a requirement to upgrade a portion of the grid 'downstream' such as beefing up a transformer in a substation hundreds of kilometers away from the wind farm.

Site Design: Once the wind study is complete, wind developers can perform a site design, laying out the wind turbines for maximum potential collection of energy from the wind. Generally, wind turbines are arranged in rows perpendicular to the prevailing wind, and spaced two to three rotor diameters apart from side to side and six to eight rotor diameters from row to row. For example, for a 1.5 MW wind turbine with a 70 meter rotor diameter, assuming the prevailing wind is from the south, the rows would line up east-to-west and the turbines would be spaced about 150 meters (500 feet) apart from east to west. The next row would be roughly 450 meters (1500 feet) to the north. This site design will usually incorporate construction drawings and an initial electrical engineering design of the site collection system (underground cables used to collect the electricity from each turbine and move the power under the gravel roads along each row to the final collection point at the on-site substation).

Equipment Assessment: Once the wind studies are complete, the developer begins analyzing how each manufacturer's wind turbine performs theoretically at the site based on their respective power curves. Each wind turbine performs somewhat differently in various types of wind resource. A small change in wind resource (and not just simple wind speed, but wind shear, turbulence and variability play major roles) can mean a large difference in energy output. Another aspect to this portion of the development process is the evaluation of the various warranty agreements and O&M (operation and maintenance) agreements available from each manufacturer. Now is the time to begin getting quotes for wind turbines, although it is unlikely that a manufacturer will provide a quote without seeing the wind data, since they will want to help recommend the height of the tower, the diameter of the rotor (length of the blades) and other variables based on the wind resource in order to maximize the performance of their machine based on the specific wind characteristics at that site. The most popular, though not necessarily the best wind turbine for every application, is the GE 1.5 MW machine. On September 30, 2004, GE Wind Energy sold its 2,500th turbine as part of ENEL's Littigheddu project in Sardinia, Italy. This makes the GE 1.5 MW unit the world's most popular megawatt class wind turbine, and shortly before this milestone GE invested $10 million in research and development to help improve technology and lower costs to increase that lead.

Financial Evaluation: With accurate wind studies and energy output projections, the wind developer can begin building financial pro-forma spreadsheets to determine possible internal rates of return, returns on cash invested by equity partners, etc.

Geotechnical Study: The foundations of these massive structures must be extremely robust due to the need for the top of the tower where the wind turbine is located to withstand enormous force from the wind (in many cases over 250,000 pounds per square inch of pressure at the hub). Therefore, the characteristics of the ground underneath where each tower will be erected must be studied. Different ground composition requires varying construction methods, depending on how much bedrock, clay, soil, or loose rock lies beneath the surface. This study also takes into account the gravel access roads that will be required.

Land Leases: Most landowners will want (and should receive) compensation for leasing their land to a wind developer. It is extremely rare for a wind developer to purchase the land. The lease generally includes an initial payment for signing a long-term (30-year or more) lease for the 'wind rights' of the land, plus royalties based on the gross kWh produced by the wind turbine(s). A rule of thumb to use is that for every MW (megawatt) of rated capacity of the wind turbine the landowner should receive slightly less than three thousand dollars per year. For example, a 1.5 MW wind turbine would typically pay the landowner about $4,000 per year. So, if a landowner had 1,000 acres and had ten (10) 1.5 MW wind turbines on their land (generally 10 is the most wind turbines you could place on 1,000 acres) they would receive about $40,000 per year for the life of the project (30 years or more). During the site evaluation and development period before construction begins, a modest good faith payment to the landowner is optional but commonplace. Because many of these sites do not ever get developed it makes sense that the landowner should receive at least something for 'tying up' the wind rights of his/her land. The land lease process should also include title searches and title insurance, as well as easement agreements, if necessary, to obtain access rights to roads and gates.

Zoning and Permitting: All necessary permits must be obtained before construction can begin. This also includes FAA permits.

Environmental Impact: Usually, either an Environmental Impact Study or Environmental Assessment must be performed (or should be, even if it is not required by local zoning regulations), depending on the zoning and permitting requirements, usually by an independent 3rd party.

Legal Review: Land leases, easement documents, warranty agreements, O&M agreements, initial financing documents, zoning and permitting documents and other documents will most likely need legal review.

3. PPA Acquisition. This step is by far the most difficult. My personal experience with this part of the process has been the most challenging experience of my life. I have been quite successful as a sales and marketing professional my entire career (even back to my high school and college years as a part-time salesperson). However, even with this track record, it is extremely difficult to get a utility to agree to buy the output of power from a wind farm. These contracts, which can be up to 100 pages in length, generally require the utility to buy the power at a fixed price as it is produced (as available), with perhaps a modest escalator in price over the term of the agreement. There are generally no penalties if the wind farm does not produce a certain amount of energy in a given day, month, or year – whatever it produces the utility agrees to buy. However, to get a PPA signed, the utility will likely require the wind developer to provide the 20-year energy forecast created as a part of the wind resource assessment process. They will want to compare the output of the wind farm, on an hourly basis, to their load (power demand) curve to determine when the energy of the wind farm will be available relative to when they need energy to meet the demand of their customers. Utility executives and system operators (grid guys, or 'sparkies') are accustomed to steady baseload, 'dispatchable,' power from coal plants, nuclear plants, and hydroelectric dams along with peaking power capacity from gas-fired power plants (available whenever it is needed, hence even these peak shaving plants are also considered 'dispatchable'). Therefore, when a wind developer attempts to have these individuals and companies to entertain buying power from an intermittent, non-dispatchable resource such as wind power, it should come as no surprise that they resist it fervently. For me, it felt like I was banging my head against a 300 foot tall brick smokestack for nearly four years before I was finally rewarded with a letter of intent from a utility. I am certain that many other wind developers feel the same way. It seems that only when public policy, such as a state mandate, requires utilities to buy clean renewable power are wind developers successful (more on public policy later in this chapter). Once a utility finally agrees in principle to buy the output of a wind project that is when the real work begins. It is time for the developer to open their checkbook because the lawyers are going to need to begin playing a major role in the process. Along with a PPA, many other documents will need to be negotiated and signed: interconnection agreement (with the utility); land lease assignments (with the landowners); warranty, O&M, purchase agreements (with the manufacturer); EPC (engineer, procure, and construct) contract (with an architectural engineering firm); insurance agreements; bank debt financing agreements; equity investor agreements; and de-commissioning agreement. All of these agreements must reference each other tying themselves together, and all agreements must meet strict requirements set forth by the financial partners who finance the project. The wind developer should expect to spend at least $500,000 in legal fees alone on this part of the process.

4. Financing. Assuming the wind developer did their homework during sections one and two of the wind development process, and assuming the energy off-taker (the utility or customer buying the power from the project) with which the PPA is executed is a creditworthy entity, then the financing should be fairly straightforward – that is, once the developer has located both debt and equity partners that are interested in financing the project. That is no small chore: at the beginning of 2003, there were only seven banks in the world that loaned money on large-scale wind farms (projects costing more than $10 Million), and none of them were U.S. banks (all were European). Since that time, some U.S. banks have entered the wind financing space, as well as Warren Buffet (through one of the subsidiaries to Berkshire Hathaway), Franklin Templeton Funds (a large mutual fund company) and several other major financial institutions. Without getting too detailed on the financing structure of wind farms, suffice it to say that these projects are generally financed using 50-70 percent debt (from a lender) at about six percent interest and the balance using an equity investor who owns the project (often in full, since they buy the entire project outright from the wind developer, and sometimes the ownership is shared) and generally expects to receive a 15-25 percent return on equity (including the benefits of any tax credits that may be available). The cost of money for the wind energy facility (the weighted average of both the debt interest rate and the ROE expected by the equity partner, usually shown as a percentage such as 12 percent) plays a significant role in the cost of energy that can be offered to the utility off-taker. Therefore, it is prudent for the developer to have initial prospective ranges for the terms of the financing established prior to even offering a proposal to the customer of the power, and certainly before final negotiations of the PPA in section three are concluded. In fact, this aspect of the wind development process should be considered ongoing, since the lender and equity investor will often want to have some influence over the nature and language of the land leases and other documents prepared at the various stages of the process.

Image courtesy Black & Veatch

5. Construction. Not to downplay the importance of construction, but this aspect is often the most straightforward of the five. Assuming the wind developer has selected an experienced and reputable EPC contractor (such as Black & Veatch, the world's largest power engineering firm), the risks of the project are mitigated in accordance with the needs of the financial partners (the wind developer often does not have the 'deep pockets' necessary to assume any major risks of the project once construction begins). Therefore, for anything that could potentially go wrong, the risk is assumed by the party responsible for that aspect of the project. For example, if a design flaw in the foundation causes the gearbox shaft to break, the engineering firm who performed the foundation design would be responsible. If the wind turbine parts began to break or the wind turbine was not performing to the standards expected according to the power curve, the manufacturer (under the warranty agreement or power availability agreement) would be responsible. And so on. The construction of a single wind turbine has three primary stages: a) pouring the foundation, which takes about 30 days to set and cure, b) the erection of the tower, wind turbine, and blades which collectively takes about 1-2 days, and c) the construction of the pad-mount transformer (the output of the wind turbine of 400-600 volts usually needs to be 'stepped up' to higher voltage to allow for lower power losses throughout the underground collection system) and other associated electrical interconnection equipment. An entire wind farm of 100 MW or more can be constructed, from start to finish, in as little as 4-12 months assuming the site development work as described in section two was all completed prior to the commencement of construction and no major delays (weather, parts availability, equipment delivery, etc.) arise.

Image courtesy Black & Veatch

Example: this exercise will describe in detail the results of installing a 1.5 MW wind turbine located in an area with average annual wind speed of 8 meters per second (about 18 miles per hour). In the interest of keeping this example simple, we'll assume that the 8 m/s wind speed equates to an average net 43 percent capacity factor.

1.5 MW – 1500 kilowatts. 1500 kW x 24 hours per day x 365 days per year x 43 percent (.43) = 5,650,200 kWh (kilowatt hours) x 3 ¢ per kWh = $169,506 gross revenue per year. Assuming a 30-year life, this wind turbine will produce $5.085 Million in gross energy revenues over its lifetime. It is also possible that the owner of the turbine could see additional financial remuneration, such as proceeds from generation capacity payments, tax credits, etc.

Image courtesy Black & Veatch

To purchase and install this wind turbine, we'll assume that it is included as part of a larger project so that the installed cost is more reasonable than installing it as a standalone turbine.   The purchase prices in 2004 of necessary major components are: 1.5 MW wind turbine ($975,000); tower and installation ($289,000); foundation ($65,000); FAA Lighting ($6,000); shipping ($65,000); pad-mount transformer and interconnection costs ($100,000); total installed cost $1,500,000. Note: if this wind turbine were installed as a standalone unit the total installed cost would range between $1.6 and $3 Million, depending on numerous factors such as crane delivery cost (or the use of a self-erecting tower), difficulties of construction at the site, and cost of components at the time.

ECONOMIC IMPACT
One of the many benefits of wind energy is the economic benefit created for the community in which the wind farm is located. This economic impact has been studied and quantified by numerous parties, and although there is disagreement (surprise, surprise) over exactly how much positive impact a wind farm may offer to a community and region, nearly all parties agree that the impact is substantial. One of my favorite studies is the Input Output Model, developed at Iowa State University (Iowa leads the United States in wind energy installed per capita, and is ranked 4th in the nation in terms of total MW of installed wind energy capacity as of mid 2004).

This model primarily includes direct, indirect, and induced economic impact benefits. It essentially says that the landowner payments represent just the tip of the iceberg in terms of total economic benefit to the region. And the conclusion of the model states that for every 10 MW of wind energy facilities installed, the total economic impact is $30 Million over 20 years. For projects that remain in place longer than 20 years (most will), the impact is even greater. Studies have also shown that property values of the land near wind farms increase over what they would have been without the wind farm (land with additional income-producing potential becomes more valuable).

According to the 2004 Windustry Project, "wind energy can diversify the economies of rural communities, add to the tax base and provide new sources of income. Wind turbines can add a new source of property value in rural areas that have a hard time attracting new industry. Each 100 MW of wind development in southwest Minnesota has generated about $1 million per year in property tax revenue and about $250,000 per year in direct lease payments to landowners. Ranchers in west Texas are welcoming the revenue from wind projects to replace declining royalty payments from soon-to-be-depleted oil wells. In addition, building wind power projects can help contribute to a stronger infrastructure of roads and power lines."

Furthermore, the Windustry study found that "every time a wind energy project is installed, it creates new jobs for people who set up and maintain the turbines. Employment opportunities range from meteorologists and surveyors to structural engineers, assembly workers, mechanics and operators. The U.S. wind industry currently directly employs more than 2,000 people, and every megawatt of new wind capacity creates 15-19 jobs and about 60 person-years of employment. The 240 MW of wind capacity installed in Iowa in 1998-1999 produced 200 six-month long construction jobs and 40 permanent operations jobs."

Finally, the good folks at Windustry remind us that wind projects support agriculture: "No one wants to live, and certainly no one wants to farm, under the spreading shadow of a nuclear or coal-fired power plant. The risk of accidents and daily exposure to [poisonous] toxins is too high. Wind turbines, on the other hand, can be installed in farm fields and pastures without ill effects on people or produce. Wind farms are spaced over a large geographic area, but their actual "footprint" covers only a small portion of the land and do not interfere with crop production or livestock grazing."

Image courtesy Vestas

PUBLIC POLICY IMPACT ON WIND DEVELOPMENT

Public policy that drives wind energy development falls primarily into two camps, the production tax credit (PTC) and the renewable portfolio standard (RPS).

The PTC is a federal tax credit that was originally signed into law in 1992. Its intent was to provide a federal incentive for renewable energy generation projects (including wind power). It works by allowing the owner of the wind project to receive a tax credit of 1.5 cents per kWh of energy produced. It is adjusted for inflation each year (in 2003 it was worth approximately 1.8 cents per kWh) and is paid for the first 10 years of the life of the project, and then ceases. The PTC usually lasts as a law for two or three years, and then expires. As long as the renewable energy project reaches what is called substantial completion before the expiration of the tax credit, it receives the PTC for the next 10 years. This is not a direct subsidy (payment from the government to the wind farm owner) but instead acts as a tax credit. In other words, assuming the wind farm owner pays federal income taxes, the PTC allows the owner of the renewable project to reduce the amount of income taxes paid by the amount of the PTC. For example, in the example above, that wind turbine would generate $84,753 in tax credits (5,650,200 kWh per year x 1.5 cents per kWh). Assuming the wind farm owner pays more than that in federal income taxes, they would be able to reduce the amount of federal income tax paid by $84,753 (and usually see a corresponding reduction in state income taxes). As a result, this can be a very valuable incentive. However, a serious problem exists with the PTC: it expires every few years and it has not been renewed on time consistently.

For example, although the PTC is said to have broad bi-partisan support in Washington D.C., it was allowed to expire after 2001 and again after 2003, forcing all U.S. wind energy projects in various stages of development to immediately cease any substantial expenditures, and causing a huge drop in the construction of wind energy projects the following year. The result has been a choppy, inconsistent growth pattern in the U.S. (see chart below):

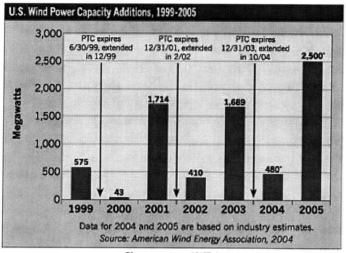

Chart courtesy AWEA

Furthermore, the PTC is extremely difficult (some say impossible) for individuals to receive any benefit from it. The way the tax code is written (section 45 of the Internal Revenue Code) for the PTC, the only parties that can truly enjoy the benefits of the PTC (monetize its value on their tax returns) are large, profitable, widely-held companies that pay a very large amount in income taxes every year, and expect to continue paying a large tax bill every year for the next 10 years (in order to fully utilize the monetary benefits of the PTC). Therefore, although this particular piece of legislation is helpful and certainly better than nothing, it does not necessarily facilitate smaller projects or ownership by individuals.

At the time of this writing (late 2004), the PTC was finally renewed nine months after it had lapsed. It was allowed by Congress to expire on December 31, 2003 and was finally renewed in October 2004 but will expire again just over a year later on December 31st of 2005. Therefore, it is extremely unlikely that very many wind farms will be built in 2004 and creates a great deal of pressure on the industry to install as much capacity as possible in 2005 before the PTC expires again. This unhealthy stop-and-start race to the finish for the wind industry creates supply bottlenecks (many turbine manufacturers are already sold out of all their manufacturing capacity for 2005 by the end of 2004) and increases prices due to high demand during the short window of PTC availability.

The second primary public policy driver is the RPS (renewable portfolio standard). The RPS is essentially a mandate which requires all utilities to purchase a portion of their electricity from renewable sources by a certain date (although it unfortunately often exempts rural electric cooperatives and city municipal electric utilities). The most common RPS standards require utilities to purchase 10 percent of their energy from renewables by the year 2015 or 2020. This can be either a federal or state mandate, but to date the federal RPS has only passed in the U.S. Senate (twice, in 2002 and 2003) but has never made it through the U.S. House of Representatives. Some states, such as California, Texas, New York, and Hawaii have aggressive RPS legislation. As of mid-2004, there are 18 states with RPS legislation. In my opinion, the state of Texas has proven to have the best RPS legislation. When it was signed into law in 1999 by then governor George W. Bush, it triggered immediate and massive action in the state, akin to a wind rush, that resulted in over $1 Billion (over 900 MW) of new wind energy generation in 2001 alone, right before the federal PTC expired. Since then, half a Billion dollars or more has been invested in the state to build new wind farms, vaulting Texas from near last to 2nd place on the list of U.S. states with the most wind energy installations.

All other forms of energy get tax credits and even outright subsidies (see next section), so the PTC and RPS should not be construed as being required for the development of wind energy projects. But they certainly do help a great deal. However, it is my hope that the PTC gets renewed only once more, for a period of five years, and then is allowed to expire permanently, never to be renewed again. In exchange for giving up the PTC, I recommend we pass a modest RPS (if we try to make it too aggressive the electric utility lobby will have a better chance at defeating it) at the Federal level, modeled after the very successful Texas RPS. This RPS should require all utilities (including rural electric cooperatives and city municipal utilities) to purchase 20 percent of their power from renewable sources by the year 2020. The RPS should have strong – and sharp – teeth in it (meaning, if the utility does not comply with the RPS, they should be penalized severely, and these penalties should be paid into a renewable energy research fund managed by the National Renewable Energy Laboratory, a U.S. Government laboratory in Colorado). Within the RPS should also be language friendly to the interconnection of renewable resources to the grid, providing fair interconnection standards that do not penalize intermittent resources.

This RPS should only require the utility to make such a purchase if they can obtain the renewable energy (or its equivalent, in the form of Renewable Energy Certificates from a renewable energy project located elsewhere in the USA) at a price not to exceed the simple average (not weighted average) of that utility's own Avoided Fuel Cost and the prior calendar year's published Avoided Fuel Cost within their RTO (Regional Transmission Operator) territory plus 30 percent, on a 20-year rolling average contract price. In other words, we need to hold the owners of renewable energy projects accountable to keeping energy costs competitive, and we already know we can meet these prices today. Plus, with a 5-year PTC and strong RPS like the one I am proposing the costs of wind and other renewable energy generation will rapidly decline even further. Utilities can purchase energy that costs more than 30 percent above their regional Avoided Fuel Cost if they so choose in order to hedge against rising and volatile fossil fuel prices, but it would not be required. This also rewards the low-cost producers of energy, since their own low Avoided Fuel Cost will lower the price threshold that renewable energy producers would have to meet in order to get a PPA from that utility.

Plus, this model encourages conversion from fossil fuels to clean, renewable energy with a 'ceiling' on the costs to society that opponents to such a conversion claim it may bring. In fact, because our current energy policy does not accurately reflect the true costs to society as described in Chapter 3 (health, environmental clean-up, the huge government subsidies that flow to the fossil fuel industry, etc.), such a public policy model could actually help stabilize and eventually reduce energy prices in the USA. Remember, what is the cost of fuel needed to run a wind farm that lasts 30-50 years?

U.S. Capitol Building

Many states would then have the option of adopting an even more aggressive RPS policy for the utilities operating within their own borders, patterned after the federal RPS. For example, since New York and California are already making progress towards 20-25 percent of their energy from renewable sources, they could accelerate their RPS policy to 30-40 percent without worrying about negative economic impacts to the state.

In conclusion, this proposed public policy model of a 5-year PTC extension and a modest but enforceable national RPS will be enough to launch wind energy into a self-sufficient, powerful industry that will no longer have – or need – any federal policy drivers. Then, wind energy, without federal assistance of any kind, will be able to compete head-to-head against coal, gas and nuclear power – all of which will undoubtedly continue to soak up and continue to require ongoing large federal subsidies just to survive.

Other public policy drivers exist, such as property tax exemptions for renewable energy projects, rebates, grants, loans, sales tax exemptions, renewable energy production incentives, net metering (this requires utilities to purchase excess power from small systems such as a residential wind turbine or solar system, in effect allowing the meter to run backwards when your home produces more energy than it needs) and public benefit funds. For more information about these and other public policy drivers and which states have implemented these drivers, there is an excellent database maintained by the Interstate Renewable Energy Council and the North Carolina Solar Center at North Carolina State University (http://www.dsireusa.org/index.cfm).

TAX CREDITS and SUBSIDIES FOR WIND

It was said that wind projects would never be built without a tax credit. Yet, wind projects are built every year all over the world, without tax credits or subsidies, simply because of the economics. Wind projects will be financed and built in the United States even if it is uncertain that a federal production tax credit (PTC) will always be in place. The absence of a federal PTC may slow the project development process as wind developers select only those sites where the energy prices they can receive are very attractive and the wind resource and construction costs are the best.

It has also been said that an industry which cannot survive on its own without federal subsidies or tax breaks shouldn't be allowed to exist. Are we to believe that the billions upon billions of U.S. taxpayer dollars flowing into the coffers of companies in the fossil fuel and nuclear industries are different from another form of a tax payer subsidy? In fact, the amount of money poured into non-renewable energy is so large relative to the pitifully small PTC that few people in government outside the Department of Energy probably know the PTC even exists! Think about it, the coal industry receives about $500 Million per year. Oil and gas exploration credits are enormous and drain federal coffers of billions and billions of dollars every year. The nuclear industry has cost U.S. taxpayers an estimated $2 Trillion over the last two decades. Without federal subsidies and the government stepping in to clean up radioactive waste, the utilities that own nuclear plants would likely have to shut them down because they are too costly to operate and maintain. There are huge federal costs to clean up the damage caused by fossil fuel infrastructure crumbling and spills and the oil industry receives major federal and state incentives to continue providing a reliable stream of fuel for our power plants and transportation sector.

Without including the direct health, environmental, and military costs (to protect our oil interests overseas) of our current energy policy, the U.S. government spends a whopping $20 Billion per year[89] to support non-renewable and dirty energy. This cost, which is actually closer to $50 billion when all government costs are included, rises every year and will equal $1 Trillion by the year 2025, represents just the direct energy subsidies and lost tax revenue from the fossil fuel and nuclear industries.

The entire wind energy industry has roughly 6,000 MW of installed generation capacity. Most of this capacity was installed prior to 2002. To make this example as conservative as possible let us assume that all 6,000 MW were built this year (2005). The following calculation will show federal tax revenue lost as a result of the PTC. 6,000 MW x an average 35 percent capacity factor x 10 years (it is actually fewer years than this, so this example overstates the estimated cost of the PTC) = 18.396 trillion kWh per year x 1.8 cents per kWh = total lost tax revenues per year of $331,128,000. $331 Million is less than what many Fortune 500 companies pay in federal income taxes to the IRS every year! And, the understated $50 billion annual cost to the government of non-renewable energy is 150 times larger than the total overstated cost of the PTC. The true cost to society of our current fossil-fuel-based energy policy is actually nearly $1 Trillion per year, as shown in Chapter 7,or 45,000 times larger than the cost of the PTC.

---

[89] Fiscal year 2005 Mid-Session Review budget of the U.S. government Executive Office of the President (July 30, 2004).

Suffice it to say that the cost to U.S. taxpayers of the PTC is but a drop in the bucket. If we are smart it will cost taxpayers nothing after a final 5-year extension if the public policy model proposed herein is enacted – and/or if we implement the exciting Freedom Plan outlined in Chapter 6.

CONCLUSION

The benefits of wind energy far outweigh any real or perceived costs and drawbacks. Wind energy's critics and skeptics need only perform a cursory amount of research to quickly learn that they have no arrows in their quiver save one: the fact that some people just do not like the way modern wind turbines look. If that is truly all the critics have to lean on as we have learned in this chapter, then it quickly becomes obvious that the continued growth of wind energy is inevitable. Where it will flourish and how fast it will grow depends on public policy drivers, the cost of alternatives to wind such as coal, gas and oil, the continued availability of oil from unstable regions of the world, and finally the most important factor, demand from the public.

Now that you have a clear understanding of how the wind energy industry works, let us explore how together we can use this newfound knowledge to accelerate the development of clean, renewable, American-made wind power to not only power an ever-increasing percentage of our electric grid but also use wind power to create abundant, safe hydrogen fuel for our cars and trucks. Turn the page to learn how America could never again have to import another drop of oil...

Chapter 5:

HYDROGEN

"The winds of the Midwest could power our entire electric grid PLUS produce enough hydrogen for every car and truck in America using a clean, reliable process called the electrolysis of water for less than 75 cents per kilogram (equivalent to a gallon of gasoline at 76 cents)."
Troy Helming, creator of The Freedom Plan

Hydrogen is the most abundant element in the universe, making up 75 percent of the known mass. It is in air, water (H_2O), oil, gasoline, natural gas, propane, soil, rocks, and even outer space. But hydrogen (H_2) is nearly always bonded to other atoms, so the bond must be broken to obtain pure H_2.

It is important to note that hydrogen is not an energy **source** like wind, oil, coal or the sun. Instead, it is an energy **carrier** like gasoline or electricity. Its value as an energy carrier is considerable because like oil and gasoline, but unlike electricity, it can be *stored* in large quantities, can be made from almost any energy *source* (including wind power), and can be used to provide almost any energy *service* (including running vehicles and electric power plants).

USES FOR HYDROGEN

The world's hydrogen industry is already quite large, producing 25 percent as much volume annually as the huge global natural gas industry. Hydrogen is used to refine oil into gasoline and other fuels (to hydrogenate fuel molecules is to improve the hydrogen-to-carbon ratio, which then results in improved octane ratings), to remove sulfur from diesel and other fossil fuels, can be used for blimps or hot air balloons (it is the lightest atom at 14.4 times lighter than air, even lighter than helium), rocket propulsion (NASA uses a great deal of H_2 to launch the space shuttle and other rockets), cooling fossil fuel power plant components, burning in an internal combustion engine (such as your current vehicle) in lieu of gasoline or diesel, and is the primary energy source for fuel cells.

HYDROGEN SOURCES

Today, nearly all hydrogen produced (over 90 percent) in the U.S. is made from natural gas, using over 5 percent of U.S. natural gas output in a costly steam-reformulation process or at chloralkali chemical plants (these plants make chlorine, a by-product of which is the 'waste' production of H2).

"Most of the hydrogen needed to displace all the world's gasoline is *already being produced* for other purposes, including making gasoline."[90]

Water, the best – and cleanest – source of Hydrogen
Image courtesy Morguefile

Another method – in my opinion the <u>only</u> one to consider – is the 'renewable hydrogen' method. This refers to production of H2 using sources that are clean and renewable, which means they would not rely on fossil fuels or other finite natural resources. Examples would be: wind, solar, geothermal, and hydroelectric. How is this done? Generally, by electrolyzing water; this involves using electricity generated from one of these renewable sources to power a device called an electrolyzer. Electrolyzers 'split' water (H_2O) into hydrogen and oxygen by passing electric current through a cathode and anode. The technique was invented in the year 1800 and has been used by NASA and countless other organizations over the years to produce hydrogen from water. This is a mature and proven technology that requires only two things: electricity and water.

The problem is that creating electricity from a renewable source usually has a higher value when sold as electricity than by converting that electricity to hydrogen. Furthermore, the loss of power through the electrolyzing process ranges from 10-40 percent or more, depending on the efficiency of the

---

[90] Twenty Hydrogen Myths by Amory B. Lovins, Rocky Mountain Institute, 9/2/03 p.8.

electrolyzer technology. Two of the largest companies in the electrolyzing business are Stuart Energy (based in Ontario, Canada) and Norsk Hydro (based in Stockholm, Sweden). These and many other companies have commercially available electrolyzers and hydrogen fueling stations today. Costs are roughly $1,000 to $2,000 per kilowatt, and many of these units can last up to 50 years or more with regular maintenance. With significant sales volume, these costs are projected to drop to $400 per kilowatt or less.

As you will learn in the next chapter, The Freedom Plan produces clean, safe hydrogen from wind power at night during "off-peak" hours when the demand for (and value of) electricity is much lower than during daylight hours. This is considered a hybrid method since it produces electricity when its value is the highest (peak hours and summertime months) and produces hydrogen by sending electricity to a bank of electrolyzers located on-site or nearby (in close proximity to a water source) during the night-time, or off peak, hours.

Using such a hybrid method allows renewable energy facilities to diversify their income stream, assuming customers are identified who are willing to purchase the renewable hydrogen. These customers might be fleets of vehicles that have converted their engines to run on H_2, homes or businesses that are powered by fuel cells, or even a hydrogen pipeline that may be used to transport large quantities of hydrogen to a distant region. The difficulty of finding such customers is where the 'chicken and egg' dilemma enters the equation, as we'll discuss later in this chapter.

Hydrogen is Clean

In the fall of 2003, the Mayor of Chicago stood behind a city bus surrounded by news reporters and television cameras. The Chicago city bus was running and liquid was steadily dripping out of the tailpipe. While on live television in front of the news cameras, he bent down, placed a glass under the tailpipe to collect some of the liquid, then stood up, raised the glass to his lips and drank deeply.

You see, that Chicago bus was powered by a Ballard fuel cell running on 100 percent hydrogen. The process of injecting hydrogen into a fuel cell produces just three outputs: electricity (which is used to power the electric motor that propels the vehicle), heat, and pure water. The process is not much different when injecting hydrogen into an internal combustion engine (ICE) such as a gasoline or diesel engine, either of which can be converted to run on hydrogen. A hydrogen-powered ICE has four primary outputs: torque (used in conjunction with a transmission to turn the wheels of the vehicle), heat, pure water, and depending on the injection calibration of the H2-powered ICE, possibly a little nitrous oxide (NOx). Because NOx are harmful to health and can cause smog, it is imperative that those mechanics performing ICE conversions take great care to adjust the injection ratios and mix of outside air to eliminate NOx emissions.

Therefore, although hydrogen critics often say hydrogen-powered cars are decades away, this is simply not how it has to be. It is true that affordable fuel cell vehicles may be decades away. But many people do not realize that we can start converting ICE vehicles to run on hydrogen today. An internal combustion engine (including the one in <u>your</u> vehicle right now) running on hydrogen creates no carbon dioxide (the leading global warming gas), no carbon monoxide (now there is no risk of accidental – or intentional – death from this deadly odorless gas), no sulfur dioxide (leading cause of acid rain) and no nitrous oxide (a leading cause of smog – assuming the conversion was performed properly). In fact, driving around in a hydrogen-powered vehicle *can actually help clean the air* in your town. This is possible because an internal combustion engine running on hydrogen requires outside air to operate.

The pollution inherent in our air today includes dust, pollen, mold, dirt, particulate matter, carbon monoxide, sulfur dioxide, countless rogue toxins and of course, carbon dioxide. This polluted air, when sucked into the combustion chamber of an H2-powered ICE burning at an extremely high temperature incinerates everything in the air instantly upon combustion. Therefore, you are literally cleaning the air as you drive around, dripping only purified water onto the street which flows harmlessly down gutters and back into the environment.

Hydrogen is Safe

As we will learn below in the Mythbuster section, hydrogen is actually quite safe. It is nearly impossible to get burned by ignited hydrogen gas, since the gas is so light it goes straight up into the air. Furthermore, hydrogen is devoid of carbon and sulfur (unlike gasoline and other fossil fuels) so it contains no harmful vapors like carbon monoxide or sulfur dioxide and when burned, it will not create toxic acids like gasoline or diesel will. One can breathe hydrogen gas in an enclosed space without risk of passing out or death. Compressed hydrogen, like nearly any compressed gas, must be handled with care, but great lengths have been taken by the decades-old hydrogen industry to dramatically mitigate the risks.

Your Car can run on Hydrogen – Today

The four-stroke internal combustion engine was originally designed to run on hydrogen, not gasoline. Its' inventor, Nicholas Otto, created the four-stroke internal combustion engine (ICE) in the early 1900s to run on 'Town Gas,' which was used at the time to light street lamps and other devices. Town Gas is comprised primarily of hydrogen, with some carbon monoxide, a little natural gas and other gases depending on where it was obtained. In fact, Otto said "I much prefer using town gas [which is 50-60 percent hydrogen] over the new waste product of the oil industry called gasoline. Gasoline is far too dangerous to use as a fuel."[91]

An internal combustion engine optimized to run on hydrogen should be 30-50 percent more efficient than today's gasoline engines. BMW hopes to raise the efficiency of its H2 internal combustion engine to a total of 50 percent, which would approach the efficiency of today's fuel cells and nearly triple the average efficiency of today's gasoline engines.

Image courtesy BMW

Nearly 200 hydrogen fueling stations exist worldwide in 2004, and that number is increasing by about 25 percent per year according to the National Hydrogen Association. Many stations contain pure hydrogen on one side of the pump, and Hythane (a blend of natural gas and hydrogen) on the other side.

Hydrogen Fueling Station

---

[91] H2 Nation "Why Wait for Fuel Cells?" by Larry Elliott, December 2003, p. 37.

Photo courtesy Stuart Energy

Any vehicle (yes, even yours) can be converted to run on hydrogen. Gasoline, diesel, natural gas or propane vehicles can all run on hydrogen. Hydrogen has greater energy content by weight than any other fuel. But this is actually more of a disadvantage, since hydrogen gas is so light that by *volume* – not weight – it actually has far less energy than gasoline or other fossil fuels. It takes a great deal of hydrogen to power an engine with the same efficiency as fossil fuels. Therefore, hydrogen must be highly compressed or liquefied to improve the energy density, and even then it still does not have nearly the energy content as gasoline does when measured by volume. Therefore, the conversion of an ICE (internal combustion engine) has some real challenges.

CHALLENGES OF A HYDROGEN ECONOMY

Hydrogen storage
By far the most difficult obstacle to tackle in converting an ICE to run on hydrogen is the storage of hydrogen fuel. In Europe, BMW and other auto manufacturers have committed themselves to liquid hydrogen. It has more energy content by volume than compressed gaseous hydrogen (at 3600 psi or less) but must be cooled to -345 degrees Celsius using energy intensive cryogenic liquification plants and then kept at that extremely low temperature in the vehicle using cryogenic tanks. In the rest of the world, a race is taking place between higher pressure gaseous hydrogen (at up to 10,000 psi, or 700 bar) and solid oxide hydrogen storage (an example is a tank made by Ovonics, a division of Energy Conversion Devices). Other methods are under development, such as carbon nanotubes, crystalline and amorphous storage. Today's ICE hydrogen conversions in the U.S. typically utilize impact-resistant, carbon fiber reinforced high pressure gaseous hydrogen storage tanks made by such companies as Quantum and Lincoln Composites. A 5,000 psi (350 bar) tank today costs about

$5,000 and a 10,000 psi tank (700 bar) will set you back over $40,000 with 2004 prices. This high cost is due to the low production volumes, but will drop significantly with economies of scale. This represents the largest single cost of the hydrogen conversion process.

Exhaust

Because of the fact that ICE's burning hydrogen emit pure water the exhaust system must be replaced with stainless steel to prevent rust.

Injection

Hydrogen has a wide temperature range of ignition capabilities (it ignites at a much lower temperature than gasoline or diesel, making it ideal for cold weather starts and can also ignite at a higher temperature). It also requires more gas (again due to its light nature and low energy density) to be injected into the combustion chamber than liquid fossil fuels. So, depending on whether your conversion is intended to make your car a Dual Fuel vehicle (capable of burning hydrogen fuel or fossil fuel depending on availability, each stored in separate tanks) or strictly a hydrogen-powered vehicle, the fuel injection system must be modified accordingly. Also, care should be taken to ensure that the air intake will not result in an abundance of nitrous oxide emissions. Performance of an ICE vehicle converted to run on H_2 should be similar to its conventional fossil fuel in most cases.

Range

Many vehicles will not be practical to convert to hydrogen, due to their weight and large, powerful engines. This would include many SUVs. A car with only a 100 kilometer (62 mile) range would not be very practical or very popular – unless it was a fleet vehicle or used solely to drive around town on short trips. To get a satisfactory range, vehicles must be light and fuel efficient to begin with, have additional hydrogen storage tanks included in the conversion process, have greater hydrogen storage capacity in those tanks (due to higher pressure), or all of the above. Designs exist for extremely lightweight vehicles (with carbon fiber frames as described in Chapter 2) that can use smaller more efficient engines to extend the range of hydrogen-powered vehicles. And when fuel cells become affordable, they will extend the range even further. This is because fuel cells are generally far more efficient (50-70 percent or greater) than ICEs (which operate at an average of only 15-17 percent efficiency). Therefore, they will require much less hydrogen than an ICE to propel the vehicle the same distance.

Hydrogen can solve the Natural Gas crisis

The United States began importing large quantities of natural gas for the first time ever in the year 2001. The largest natural gas field in the world (in Hugoton, Kansas) is expected to run dry between the years 2012 to 2015. As discussed further in later chapters, natural gas shortages, price volatility and increasing cost will place huge burdens on the U.S. economy in the next two decades.

Natural gas is generally comprised of at least 95 percent methane (CH_4), the rest being ethane and propane. Methane has one carbon atom and four hydrogen atoms in every molecule, which means methane is comprised primarily of hydrogen. The delivery of natural gas is accomplished through a sophisticated pipeline system and with LNG (liquefied natural gas) tankers.

The bulk of the natural gas pipeline infrastructure exists in two primary regions: the Gulf of Mexico and the High Plains area (western Kansas/Oklahoma and the Texas panhandle). It is noteworthy that the High Plains area also happens to be where some of the best wind resource in the world is located. It should come as no surprise that The Freedom Plan described in the next chapter takes full advantage of this unique coincidence. Or, perhaps there is no such thing as a coincidence and it was meant for this area to have the greatest wind resource and ample existing natural gas pipelines. Either way, the fact remains that significant gas infrastructure already exists in exactly the area where the wind resource is abundant. Let us explore why this is significant.

Because natural gas is comprised primarily of hydrogen (70-75 percent), anything that runs on natural gas can run on hydrogen – although some natural-gas-powered devices may require minor modification. For instance, a modern home heating furnace could run on hydrogen. Why would one want to do that? Because burning hydrogen in your furnace burns more cleanly with zero carbon dioxide emissions, produces zero deadly carbon monoxide, and creates pure water as a by-product that could be recaptured as drinking water if desired. And most importantly, it does not deplete a scarce natural resource that can be very costly with an unpredictable price.

Hythane

Because the ignition properties of hydrogen differ from those of natural gas (and because of other reasons), burning 100 percent hydrogen in a residential furnace, a gas-fired turbine in a power plant, or a natural-gas-powered vehicle may not be the ideal solution. A better option may be to blend hydrogen with natural gas – this blended fuel is called Hythane. Nearly anything that runs on natural gas can run on Hythane without modification, as long as the blend of hydrogen does not exceed a certain composition threshold (usually no more than 10-20 percent hydrogen).

Benefits of switching to Hythane instead of Natural Gas:

Mitigate the natural gas crisis – even a 10 percent reduction in natural gas use by blending 10 percent renewably produced hydrogen can significantly impact natural gas supplies. This would partially stabilize both the price and the storage supplies of natural gas and create a much needed new market for renewably produced hydrogen, providing increased competition to the natural gas marketplace and further mitigating prices.

Lower emissions – according to studies performed by HCI (Hydrogen Components Inc. of Littleton, Colorado) as part of a Hythane bus fleet used at SunLine Transit in California, adding a mere 5 percent hydrogen to the natural gas supply lowers emissions of poisonous carbon monoxide and smog-forming nitrogen oxide by a whopping 50 percent.

Cleans the air – the use of Hythane (especially in blends with higher proportions of hydrogen) in combined-cycle natural gas turbines in power plants can actually clean the air on a large scale. As air from the town or city where the power plant is located is sucked into the turbine, the higher hydrogen concentration can actually burn out the carbon dioxide, sulfur dioxide, particulate matter, dust, pollen, mold and other pollutants in the air – literally cleaning the air in your town whenever Hythane and/or hydrogen is burned at the power plant.

Cleans the water – presently, nearly all power plants, internal combustion engines, furnaces and other devices using natural gas create water that is contaminated with formaldehyde, sulfur, and other toxins. Burning hydrogen creates crystal clean drinking water, so it should go without saying that replacing natural gas with Hythane produces cleaner water that flows back into the environment. Similar to the air emissions of Hythane, a small percentage of hydrogen has a much larger percentage reduction in water pollution.

Hydrogen Delivery via Pipeline

The similarities between natural gas and hydrogen would seem to suggest that using existing natural gas pipelines to transport large quantities of hydrogen would be just fine. For the most part, this is true. There are, however, two challenges of hydrogen delivery via pipeline that need to be addressed: embrittlement and storage.

Critics of the hydrogen economy claim that moving hydrogen through pipelines will cause the steel of the pipe to become brittle, eventually weakening enough to cause the pipe to crack or break. They claim that because the hydrogen atom by itself (H) and molecular hydrogen (H_2, which is two hydrogen atoms bonded together) are so small, the size allows the tiny atoms to seep into metals and seals, ultimately escaping through the metal itself. This is simply not true. Why it is not true will be addressed here instead of the Mythbuster section since it is not a widely held myth.

Image courtesy Stuart Energy

According to the book "The Solar Hydrogen Civilization" by veteran Arizona State professor, renowned scientist, and founder of the Hydrogen Association (in 1964), otherwise known as "Dr. Hydrogen" Roy McAllister, "hydrogen has been stored at 2,000 PSI since the early 1900's in ordinary steel cylinders, which show no sign of degradation due to long-term exposure to pure hydrogen. Ordinary engines converted to operation on hydrogen show no sign of metal embrittlement or other degradation after fifteen years of pollution-free service. Engine oil stays clean, spark plugs last much longer, and degradation as measured by corrosion and wear on ring and bearings is greatly reduced." In fact, one cylinder used to store hydrogen at 2,000 PSI in the fuel cell laboratory at the University of Kansas was stamped with a hydrostatic test date of 1907. It has been used and reused, again and again, to fill and discharge hydrogen for nearly 100 years, with a safety test date stamped on the neck of the cylinder from 1907 and every five years thereafter. Chemical analysis of metal filings from this cylinder were taken and found it to contain 99 percent iron, 0.6 percent manganese, 0.2 percent carbon and a few other elements. In other words, it contains not even as much alloy as steel from a standard farm plow does – and no liner was present in the cylinder either. This and many other tests have shown that simple "mild" steel containers hold hydrogen just fine, without leakage or embrittlement.

Therefore, it follows that hydrogen should be able to flow through pipelines without leaking and without causing embrittlement. Unfortunately, that is not always the case. With changing pressures of hydrogen in the pipe, the pipeline's seals may eventually see significant degradation and ultimately fail. This is due to 'rogue hydrogen' in the pipeline seals. The welding process used to connect the pipeline sections can introduce a form of hydrogen through use of damp welding rod flux or welding in a wet environment. Arc welding electrolyzes any water present in the process into hydrogen and oxygen ions. Ionic or nascent hydrogen particles are much smaller than standard diatomic hydrogen molecules. Master welders know how to avoid hydrogen embrittlement: they keep their welding rods dry (storing them in sealed canisters) and prepare their work area – making sure it is clean, preheated, and all pieces are kept dry. This avoids the embrittlement problem by preventing rogue hydrogen introduction into the steel.

About 1,500 km (930 miles) of special hydrogen pipelines exist today in many parts of the world and roughly one-half exist in North America alone.[92] Most of them, however, run at constant temperature and pressure (up to 100 bar). Hydrogen flows better than natural gas but is less dense and takes more compressor energy (assuming compressors are needed, which may not be the case if H_2 production is accomplished with high-pressure electrolyzers). Therefore, transporting hydrogen through existing natural gas pipelines will likely have energy losses of about 20-25 percent, compared to natural gas, including losses for compressors[93] and only about 10-15 percent energy losses if compressors are not required. Pipelines covering large distances may require special polymer liners (like the ones currently used to renovate old water and sewer pipes), adding a hydrogen-blocking metallized liner or coating and/or construction using carbon composites to accommodate varying pressures of hydrogen, but it can certainly be done. In fact, a 200-mile crude oil pipeline has already been converted to carry hydrogen.[94] And many older pipelines are

---

[92] "Hydrogen Pipelines," HyWeb, 12/18/02, Ludwig-Bölkow-Systemtechnik GmbH LBST German Hydrogen Assoc., www.hydrogen.org/News/arcv402e.html#LBST%20Analysis%2002-12-18, updated by B. Kruse *et al.* (ref. 14), p. 28.

[93] Air Liquide, "Hydrogen Delivery Technologies & Systems," *Procs. H2 Delivery Workshop,*www.eere.energy.gov/hydrogenandfuelcells/hydrogen/wkshp_h2_delivery.html, 7–8 May 2003, Sandia National Laboratories.

[94] W.C. Leighty, M. Hirata, K. O'Hashi, H. Asahi, J. Benoit, & G. Keith, "Large Renewables–Hydrogen Energy Systems: Gathering & Transmission Pipelines for Windpower & other Diffuse, Dispersed Sources," *World Gas Conf.* 2003 (Tokyo), 1–5 June.

already hydrogen-ready since they were originally designed to carry "town gas" (synthetic coal gas that is 50-60 percent hydrogen by volume).

Finally, the storage of hydrogen on a large scale is another challenge to be solved. Today, very large quantities of hydrogen are being stored in enormous wet salt caverns in Europe. It is likely, though untested, that hydrogen can also be stored in large quantities in depleted natural gas and oil fields.

For example, as mentioned earlier some of the best wind resource in the world is in the High Plains region of western Kansas and Oklahoma, eastern Colorado and the Texas Panhandle. The High Plains area also contains the world's largest natural gas field – when this field was full it held enough natural gas to theoretically meet 100 percent of U.S. natural gas demand for years. The Hugoton gas fields are now running dry. There is also an abundance of 'graywater' in the region that is unsuitable for farming. This area also has a huge network of existing natural gas pipelines, second only to the Gulf Coast region. So, one can get excited thinking about the possibilities of using wind and water to create hydrogen through electrolysis, blending that hydrogen into natural gas to create Hythane to help solve the natural gas crisis, using some of those pipelines for pure hydrogen delivery, and storing huge quantities of renewable hydrogen by injecting it back down into the Hugoton gas field. This process effectively stores the wind energy in the form of hydrogen which can be used whenever it is needed. It will likely have another fringe benefit: by injecting hydrogen into the Hugoton field it should squeeze the last remaining natural gas out of the field and extend its useful life. One might call this a win, win, wind solution.

Will renewable Hydrogen production use up too much water?

The use of hydrogen in engines (or fuel cells) returns the water that was used to create the hydrogen in the first place. In fact, the full circle of hydrogen use, from renewable production, to storage and transportation, to consumption (in an internal combustion engine or fuel cell), to the by-product of consumption (pure water vapor in the exhaust from an engine and/or the water produced by fuel cells) is a Net-Zero Impact.

One gallon of gasoline weighs about 6.4 pounds; when combusted it produces 19.76 pounds of carbon dioxide and 9.09 pounds of water.[95] A gallon of water weighs about 8.33 pounds. Therefore, burning one gallon of gasoline produces <u>more</u> than a gallon of water which passes through the tailpipe, primarily in

---

[95] Solar Hydrogen Civilization, page 173.

vaporous form. And, this is not water one would want to drink. Burning fossil fuels throughout the world adds 190 million barrels of water to the Earth's surface water every day, most of it polluted with carbonic acid that forms as carbon dioxide is absorbed into the water dripping from automobile tailpipes. This may be contributing to lowering the salinity of ocean water, which has been theorized to cause any number of adverse affects, including more violent Atlantic hurricanes, a slowdown of the ocean thermal conveyor system, accelerated release of methane stored deep in the ocean in the form of methane hydrates, and others.

Niagara Falls (image courtesy http://www.forsakenxlies.deviantart.com)

Replacing a gallon of gasoline can be accomplished with just two pounds of hydrogen which can be produced from 18 pounds (just over two gallons) of water.[96] Burning this hydrogen in your vehicle's converted internal combustion engine (or a fuel cell) will return the 18 pounds of water that was used to produce it, meaning the net effect on water equals zero. And, the water returned through this process is not contaminated with carbon or sulfur.

In other words, all the water used up in the creation of renewable hydrogen is returned – in pristine form – back into the environment. In some cases the water source for the production of renewable hydrogen may be 'brownwater' or 'graywater' (municipal wastewater or other non-potable water source) which is filtered prior to processing by the electrolyzer, and is then returned as pure, clean, drinkable water back into the environment. And, none of this pure water

[96] Solar Hydrogen Civilization, page 173.

is 'extra' water, since it all came from water in the first place. Using hydrogen instead of fossil fuels will no longer contaminate hundreds of billions of gallons of water per year. The gain if all vehicles were running on hydrogen would be 17 billion gallons of crystal clear drinkable water returning to the Earth[97] - every single day.

10 HYDROGEN MYTHS - BUSTED

Common hydrogen myths are listed and then methodically dispelled.[98]

Myth #1: An entire hydrogen industry would have to be developed from scratch, which would take 30+ years. As discussed already in this chapter, hydrogen can be produced from wind or solar energy (and water) where it is most practical and economic. Then, it can be injected into natural gas to form Hythane to incrementally begin to build the hydrogen industry. As production and demand levels grow, pure hydrogen pipelines will inevitably be dedicated to widespread delivery of hydrogen to fueling stations that can be added quickly at existing gas station locations in major metropolitan areas. From these fueling stations local hydrogen delivery to office parks and neighborhoods using fuel cells is likely to occur. This will be due to simple market demand and economics, assuming the delivery trucks do not have to transport the hydrogen more than 50 kilometers – 31 miles – from the fueling station. Community hydrogen pipeline delivery systems will take a little longer, but can be developed based on demand. Finally, homeowners with the financial wherewithal and desire to make their homes "Energy Independent" will purchase Home Hydrogen Systems (described in later chapters) that produce their own hydrogen onsite from rainwater and wind or solar power. This hydrogen would then be used to power their home with a fuel cell and would include a fueling appliance in their garage to fill up their vehicles with H_2 made at their own home. The entire nationwide conversion to hydrogen could theoretically take place in five years, but The Freedom Plan calls for a more realistic timeframe of 10 years in order to fully develop the hydrogen infrastructure – much shorter than 30 years and certainly not built from scratch. The next chapter describes the hydrogen infrastructure build-out portion of The Freedom Plan.

Myth #2: Hydrogen is too dangerous, explosive, or volatile for use as a fuel. The Hydrogen industry has an impressive safety record dating back over 50

---

[97] Solar Hydrogen Civilization, page 174.
[98] Portions of this section have been reprinted with permission from Twenty Hydrogen Myths by Amory B. Lovins, Rocky Mountain Institute, 9/2/03.

years. Although all fuels have inherent risks, those of hydrogen have been extensively proven to be easily mitigated with proper care and handling.[99]

The technical characteristics of hydrogen include: 1) It is extremely buoyant (14.4 times lighter than air compared to natural gas at 1.7 times lighter than air); 2) It is highly diffusive (four times more so than natural gas and 12 times more so than gasoline vapors), so leaking hydrogen rapidly rises upward, away from the source of the leak;[100] 3) It burns seven percent cooler than gasoline (when ignited, hydrogen burns very quickly with a non-luminous flame that cannot easily scorch anything at a distance); 4) It emits only one-tenth the radiant heat of a gasoline or other hydrocarbon fire;[101] and 5) It produces absolutely no smoke, which according to firefighters is the number one killer in fires.

Unlike natural gas, leaking hydrogen exposed to a spark is far likelier to burn than to explode, even inside a building or a car, because it burns at concentrations far below its lower explosive limit. Leaking hydrogen will nearly always burn and not explode when lit. And in the rare case that it might explode, the theoretical explosive power per unit volume of gas is 22 times *weaker* than gasoline vapor.[102]

The 1937 Hindenburg disaster created a fear of hydrogen that is completely unwarranted: the safe characteristics of hydrogen gas actually <u>saved</u> 62 lives! NASA scientist Dr. Addison Bain determined that the loss of life would have been identical if nonflammable helium had been used by the dirigible and that not a single person was killed by the hydrogen fire.[103] Those who died (35 percent of the passengers) were killed by jumping out, by the burning cloth canopy (which was covered by what would be called rocket fuel today), by burning diesel oil that was used to power the propellers, or by falling debris. The survivors (65 percent of the passengers) simply rode the burning dirigible to the ground as clear hydrogen flames whirled harmlessly up into the air above.

---

[99] W.C. Leighty, M. Hirata, K. O'Hashi, H. Asahi, J. Benoit, & G. Keith, "Large Renewables–H2 Energy Systems: Gathering & Transmission Pipelines for Windpower & other Diffuse, Dispersed Sources," World Gas Conf. 2003 (Tokyo), 1–5 June.

[100] Air Liquide, "Hydrogen Delivery Technologies & Systems," *Procs. H2 Delivery Workshop,* www.eere.energy.gov/hydrogenandfuelcells/hydrogen/wkshp_h2_delivery.html, 7–8 May 2003, Sandia National Laboratories

[101] Twenty Hydrogen Myths by Amory B. Lovins, Rocky Mountain Institute, 9/2/03 p.9

[102] A.D. Robinson, "Hydrogen Hype," *Access to Energy* 30(9):1 (April 2003) and Twenty Hydrogen Myths by Amory B. Lovins, Rocky Mountain Institute, 9/2/03 p.9.

[103] A. Bain & W.D. Van Vorst, "The *Hindenburg* tragedy revisited: the fatal flaw found," *Intl. J. Hydr. En.* 24:399–403 (1999); A. Bain & U. Schmidchen, "Afterglow of a Myth: Why and how the 'Hindenburg' burnt," www.dwv-info.de/pm/hindbg/hbe.htm.

The Hindenburg, 1936

If hydrogen gas were to escape into a closed room such as your garage, it would accumulate near the ceiling. Gasoline vapors or propane in that same garage will accumulate near the floor (natural gas, being only slightly lighter than air, will remain mixed with the air for some time until the point when most of will eventually concentrate near the ceiling). Standing in a carpet of fire is obviously far more dangerous than standing below a non-explosive, non-luminous flame that rises upwards when burning and emits little radiant heat.

A recent government test was performed to compare a gasoline fire with a compressed hydrogen gas fire inside a vehicle. First, a hydrogen leak was created (a very unlikely event requiring a triple failure of hydrogen industry standard redundant protection devices). The leak was caused intentionally at the highest-pressure location and discharged the full H_2 storage of 1.54 kg of the car in about 100 seconds. The flame caused the passenger compartment temperature to rise by, at most, only about 1-2 ºC (0.6-1.1 ºF). The highest temperature readings nearest the vertical flame were no higher than what would be found near a dashboard while a car is sitting in the sun. And, the interior of the vehicle was unscathed, sustaining no damage to the surrounding fabric and plastic. In the second test, a leak from a 1.6 mm (1/16") hole in a gasoline fuel line had 2.5 times less energy escaping from the leak and yet it gutted the car's interior and would have certainly killed anyone trapped inside. Remarkably, because the hydrogen leak didn't damage the car, both tests were conducted *using the same car*.[104]

Myth #3: Making hydrogen uses more energy than it yields, so it is too inefficient. Making hydrogen from natural gas reformers is 72-85 percent

---

[104] M.R. Swain, "Fuel Leak Simulation," www.eren.doe.gov; C.E. Thomas, personal communication to Amory Lovins of RMI, 4 June 2003; Twenty Hydrogen Myths by Amory B. Lovins, Rocky Mountain Institute, 9/2/03 p.10.

efficient[105] and 70-85 percent efficient[106] in water electrolyzers. Therefore yes, of course there are energy losses making hydrogen. However, making electricity at a coal plant is only 15-45 percent efficient. Reforming oil into gasoline, transmitting electricity long distances over power lines, and pumping natural gas through pipelines all have energy losses as well. But hydrogen's greater end-use efficiency actually enhances the losses resulting from hydrogen production, making hydrogen one of the most attractive round trip fuels available. Remember, the energy used to make hydrogen in The Freedom Plan is inexpensive wind power during off-peak (nighttime and spring/autumn) hours. When the lights are off at night (and hence electric demand is low), the wind keeps blowing. With hundreds of thousands of megawatts of wind energy installed in the USA, without a demand load to send this power to, wind farms would have to be curtailed (taken offline) to prevent overloading the system. But what some see as reason to *not* build huge quantities of wind farms, I see as an *opportunity*: install thousands of megawatts of electrolyzers near water sources to accept that 'excess' wind energy that would have otherwise been wasted. Remember, all of the water used in this process is returned to the earth as pure water once the hydrogen is used as an energy source. The electric utility industry is plagued by transmission constraints, the need for costly peak power generators, and growing loads in areas where the system is already strained. Therefore, the use of hydrogen as a way to store electricity adds significant value to the economics of hydrogen production. Although the round trip efficiency of electrolyzing water to make hydrogen, storing it, and then converting it back to electricity in a fuel cell is only about 50-60 percent (still much higher than a modern baseload fossil-fired power plant) after electrolyzer and fuel cell losses, it can be used to displace electricity generating at older peaking plants operating at efficiencies as low as 15-20 percent. And, the potential revenue of selling stored power during peak hours vastly exceeds all the losses of the inexpensive off-peak power used to make the hydrogen in the first place. This and its potential use as a transportation fuel help explain why The Freedom Plan relies so heavily on hydrogen production and storage to solve our myopic dependence on foreign oil and fossil fuels.

---

[105] Boeing's exothermic One-Step Hydrogen (BOSH) process, now in testing, is predicted to be even more efficient, and to cost half as much as traditional reformers. Other developers are on similar trails; and Twenty Hydrogen Myths by Amory B. Lovins, Rocky Mountain Institute, 9/2/03 p.10

[106] Or ~80–90+% for electrolyzers using hydrogen's Higher Heating Value. R. Wurster & W. Zittel (LBST), "Hydrogen Energy," *Procs. Workshop on Energy Technologies to Reduce CO2 Emissions in Europe: Prospects, Competitio ,Synergy,* Energieonderzoek Centrum Nederland (ECN), Petten, 11–12 April 1994, www.hydrogen.org/Knowledge/ECN-h2a.html. At low load, which correspondingly decreases asset utilization, the HHV efficiency can be much higher yet: for example, Hans Hoffman of the German firm GHW (pers. comm., 3/17/03) reports 82% (HHV) at full load, rising to 98% at 1/6 load, measured for the 450-kW, 30-bar alkaline electrolyzer at the Munich airport.

Myth #4: Delivering hydrogen to end users would consume most of the energy it contains. It is true that delivering hydrogen in trucks or train cars almost never makes economic sense. The energy density of hydrogen is simply too low, meaning it must be transported either at 350-700 bar (5,000-10,000 psi) which is a very high pressure (requiring costly compressors unless high pressure electrolyzers are used) or as a liquid (requiring an extremely expensive liquification plant). Even by using one of these options, a delivery service cannot cost-justify using tube trucks or cryogenic trailers to transport hydrogen farther than perhaps 50-300 km (30-180 miles). The weight of the truck causes the vehicle to use up more energy driving it more than 50-300 km than all the energy contained in the hydrogen, regardless of the fuel used by the truck. Therefore, production of hydrogen using wind energy must take place near where it will be used so that delivery distances are minimized. And, hydrogen must be stored in cost-effective natural geologic formations, such as depleted gas fields, oil fields, wet salt caverns, underneath aquifers, or modern gasometers (large cylindrical tanks with a balloon-like high strength polymer bladder and a very heavy lid like the ones that used to store town gas). Finally, the best and most likely way hydrogen will be delivered is through existing or modified natural gas pipelines, or new dedicated hydrogen pipelines (see next myth). Nearly all large office parks and industrial centers – as well as many neighborhoods – have existing natural gas pipelines. Buildings in the U.S. use a whopping two-thirds of all electricity, suggesting that installing large fuel cells in buildings for primary or back-up power (supplied by hydrogen from its existing pipeline) will be a key market to help launch the hydrogen economy. Using pipelines to transport renewable hydrogen is inexpensive, safe, and can leverage the vast existing network of natural gas pipelines, may of which run through the best wind areas.

Myth #5: Hydrogen cannot be distributed in existing pipelines, requiring costly new ones. This myth has been clearly and methodically dispelled earlier in this chapter.

Myth #6: We cannot effectively run cars on hydrogen or store hydrogen onboard. Dozens of cars are already running on hydrogen today all over the world in numerous climates. The fuel cell process, invented in 1839, produces electricity in an electrochemical reaction with no combustion, makes pure hot water (which could be routed to a coffee maker in the dashboard or cooled for pure drinking water) and has zero pollution. Fuel cells have been used successfully by NASA in the Apollo program and in space shuttles for decades, demonstrated in a passenger vehicle by GM in 1966,[107] and have been widely

---

[107] General Motors powered the Electrovan in 1966.

used by the military in submarines and countless other applications due to their safety record, reliability and simplicity. But, storing enough hydrogen to propel today's heavy steel cars would require larger tanks at very high pressure (350-700 bar or more) and would reduce the driving range. Companies such as Lincoln Composites, Dynetek, and Quantum (partly owned by GM) sell filament-wound carbon fiber tanks lined with an aluminized polyester bladder which have up to 13 times the performance of steel or aluminum tanks, cannot corrode and are extremely strong and safe. In crashes that flatten steel cars and shred today's flimsy gasoline tanks, these sophisticated tanks were completely undamaged. The high pressure tanks are not a significant safety concern: their placement low and away from the passenger compartment would require an impact so severe that if the tank were to be damaged, no one would have survived the crash anyway. And even then, as we learned in Myth #2, hydrogen is safer than gasoline. As an example, a 700 bar impact resistant tank might contain 4-6 kg of hydrogen depending on the size of the tank. In a car getting 20 MPG today with gasoline, that car would have a 200 mile range with a 10 gallon gas tank. Since a kg of H2 is roughly equivalent to a gallon of gas, a tank with 5 kg of H2 would provide half that: only 100 miles of driving range. Today, 700 bar tanks (10,000 psi) are not yet approved for highway driving by the U.S. Department of Transportation so the actual range with a 350 bar tank would be even less. A 700 bar tank costs $40,000 since they are nearly all made by hand today. Therefore, to increase the range we must either create inexpensive 700 bar tanks in mass production (or even 1,655 bar, 24,000 psi, tanks proven by GM and legally approved for vehicular use in Germany) or lower the gross weight of the average hydrogen powered vehicle. These tanks could be mass produced for a few hundred dollars. A series of designs exist for low drag, ultralight vehicles that could be mass produced at a cost of $15,000-50,000 each depending on options. These vehicles, if powered by a fuel cell, could get 100 miles per kg of hydrogen (roughly equivalent to 100 MPG) and have a 500 mile range with a 5 kg 700 bar tank or 1,000 mile range on a 10 kg 1400 bar tank and have a 200,000 mile standard warranty. And their strength can absorb up to five times as much crash energy per pound as steel and crush more smoothly, using the crush length up to twice as effectively as today's vehicles. In closing, further research and development on hydrogen storage technologies would be nice but is not essential for rapid deployment of widespread hydrogen-powered vehicles.[108]

Myth #7: Oil, car, and utility companies will vigorously oppose the hydrogen economy as a threat to their profit and way of life. Truthfully, this myth has concerned me for years; it is the only one that has. Some people claim that the

---

[108] Twenty Hydrogen Myths by Amory B. Lovins, Rocky Mountain Institute, 9/2/03 p.17.

hydrogen development efforts of the oil and gas companies are mere window-dressing. I too thought that may be true. But after attending the National Hydrogen Convention in Washington, D.C. in 2003, I learned otherwise. The three largest booths there were ChevronTexeco, Shell, and BP. A Shell Oil director predicted in 2000 that half of all new cars will run on hydrogen by 2010. Further due diligence uncovered the fact that these publicly traded companies have each invested at least one billion dollars on hydrogen efforts. Although $1 Billion is a small percentage of gross revenues for these huge companies, it is a relatively significant percentage of gross profits and is not a small number. Furthermore, I learned how important hydrogen is to the refining process of gasoline, so these companies are already required to produce large quantities of hydrogen. General Motors has over 600 employees in Detroit working full-time on their hydrogen car programs, and all the major automakers (GM, Toyota, Ford, DaimlerChrysler, Honda, BMW, and more) have hydrogen programs and existing hydrogen-powered prototype vehicles operating in various areas of the world, undergoing real-world testing and evaluation. So yes, hydrogen is a big business for big companies. But I also learned that none of them have fully embraced hydrogen as *the* fuel of the future. Furthermore, there is some resistance to any rapid change or radical new thinking; not very surprising for any large entity which usually resists change. But until hydrogen fueling stations become commonplace the car companies will not provide us with production models of hydrogen-powered ICE or fuel cell vehicles. And until there are cars to fill, the oil companies will not spend the money to upgrade fueling stations to include hydrogen. This frustrating little dance could continue forever. Fortunately, progress is being made: California Governor Arnold Schwartzenegger signed Executive Order S-7-04 on April 20, 2004 which requires hydrogen fueling stations every 20 miles on every California interstate highway. This historic act, called the Hydrogen Highway Initiative, will forever change the way the world views transportation fuel. Will other states follow suit? Doubtfully, but some may. Fortunately, it will force those two dance partners, the oil and car industries, to finally embrace each other. But will it be enough? No, but it is a great start. Thanks Arnold, for your vision and courage to force positive change. Only through public demand all over America can we truly force our oil-based transportation industry to break the chains of dependence and soar freely on the winds of renewable hydrogen. Finally, utility company executives vehemently oppose a massive shift to renewable energy and hydrogen production. Less prone to be in touch with consumer desires, monopolistic electric utilities are extremely slow to change and stubbornly set in their ways. Yet wind power is growing steadily in the U.S., and with more demand from the public it can grow much faster. Electric utilities that embrace wind power can supply the oil industry with inexpensive off-peak energy to allow existing oil refineries to become merchant hydrogen production plants.

Myth #8: The hydrogen economy would damage the atmosphere due to leakage. According to the Rocky Mountain Institute, "switching from today's fossil-fuel economy to an all-hydrogen economy with only a one percent leakage rate would release about as much molecular hydrogen as is now released by fossil-fuel combustion, so as a first approximation, nothing would change."[109] This one percent rate of leakage is actually much higher than what has been observed for decades by the existing hydrogen industry. And the existing industry is no drop in the bucket: if all current worldwide production of H2 (roughly 50 million T/year) were used to fuel lightweight fuel cell vehicles it would displace 2/3 of today's entire worldwide consumption of gasoline.[110] Additionally, the hydrogen industry is doubling every 11 years (due to a six percent annual growth rate). Finally, the entire German hydrogen infrastructure (one of the largest in the world) loses only 0.1 percent from leakage.[111] This suggests that a one percent leakage estimate, being 10 times greater than the' statement above, is quite conservative. In reality, the actual leakage rate should be less than a mere couple tenths of a percentage.

Image courtesy NASA JSC Digital Image Collection

---

[109] Twenty Hydrogen Myths by Amory B. Lovins, Rocky Mountain Institute, 9/2/03 p.30.
[110] 50 MT H2/y at LHV (120 MJ/kg) is 6 EJ/y. In quintupled-equivalent-efficiency vehicles, that displaces 30 EJ/y of gasoline-equivalent. World consumption of gasoline in 2000 was 19.76 Mbbl/d or ~42 EJ (www.eia.doe.gov/emeu/iea/table35.html), and 30/42 is 0.71.
[111] W. Zittel & M. Altmann, "Molecular Hydrogen And Water Vapour Emissions In A Global Hydrogen Energy Economy," *Procs. 11th World Hydr. En. Conf.* (Stuttgart, 1996), www.hydrogen.org/knowledge/vapor.html.

Myth #9: **The hydrogen transition requires a big $300 billion Federal government initiative, like the Apollo Program or the Manhattan Project.** So what? If it did cost that much (it won't), that is less than the cost of the current military conflict in the Middle East.[112] Actually, the many organizations who have done more than just a passing study on the subject (such as the Rocky Mountain Institute) have determined that the total cost of a hydrogen transition is probably far less than $100-300 Billion.[113] The Freedom Plan includes a generous total cost of $250 Billion for all necessary hydrogen infrastructure components (see the next chapter for details), a large portion of which are intended not for the transportation sector but for the electric utility sector (to convert gas-fired power plants to burn hydrogen). Therefore, the numbers and assumptions behind The Freedom Plan are higher than what is really necessary, maintaining the conservative and realistic theme of The Freedom Plan's massive transformation of our energy industry.

Myth #10: Hydrogen is expensive compared to gasoline. Today, that may be the case when considering all the challenges discussed herein: lack of existing spare capacity in the existing hydrogen production industry, costs of transportation and storage of hydrogen, the perceived (but erroneous) high cost of electricity from renewable energy sources, and smaller overall economies of scale of the existing hydrogen production as compared to gasoline refining all contribute to suggest that a kg of hydrogen costs more than a gallon of gasoline. The Freedom Plan creates hydrogen at a cost *below* $1.00 per kilogram (about the same as a gallon of gasoline). This very attractive price will remain stable and can even decline gradually for 50 years.

How can hydrogen prices remain stable? Today, the price of gasoline fluctuates primarily due to the cost of crude oil. In October 2004, oil hit $50 per barrel for the first time in history, setting an all time record high. The last time oil was as high (when adjusted for inflation) was 1981, which was caused in large part by three factors:

1. Two oil embargos had occurred recently in the 1970s, which made the oil markets jittery and nervous.

---

[112] According to reports by the non-partisan Congressional Budget Office (CBO) and Government Accounting Office (GAO), the direct military cost of the conflicts in Afghanistan & Iraq launched in 2003 will cost **$300-500 Billion** through 2006. Many policy experts claim these figures are understated since they do not include indirect costs (State Department funds, hidden military & intelligence costs, etc.).

[113] Twenty Hydrogen Myths by Amory B. Lovins, Rocky Mountain Institute, 9/2/03 p.36.

2. A drastic cut in output by OPEC (Organization of Petroleum Exporting Countries) which restricted global supplies of oil which naturally increases prices.

3. The U.S. economy relied more heavily on energy as a percentage of GDP (today the U.S. economy has a higher percentage of services – as opposed to manufacturing – which do not rely as much on energy).

Today's record high oil prices exist when OPEC is producing oil at maximum capacity (the spigot is wide open, with no embargo or any other curtailment of production). Yet, the world's thirst for oil continues to increase at record pace, led primarily by the United States (with only 4 percent of the world's population the U.S. consumes 24 percent of the world's oil[114]), China (expected to equal or surpass America's oil consumption by 2020), and India. Oil is depended upon not only for gasoline and diesel to power the transportation sector (cars, trucks, airplanes, trains, and ships) but is also the sole source of polymers for plastics (toys, cars, computers, appliances, pharmaceuticals, packaging, bottles, shoes, and clothing nearly all use polymers in one form or another). The threat of supply disruptions in Nigeria, Iraq, Venezuela and Russia keep the oil futures market skittish and there seems to be an ever present fear that just a single 'dirty bomb' in Saudi Arabia, the world's largest oil producer, could instantly cripple the world's economy.

Compare that helpless dependence on largely uncontrollable factors such as global economics and the prevention of terrorist acts to maintain stability of oil prices to the relatively mundane chore of producing renewable hydrogen. Think about it: gasoline is dependent on a volatile and limited natural resource, oil, whose price can change at the drop of a hat and what's left is located primarily in unstable regions. Hydrogen produced from wind and water, on the other hand, is dependent on just two things: access to water and continued winds. Since the Earth is nearly two-thirds covered with water, finding it or running out will never be a problem. And, if the wind ever stops blowing, we have much more serious problems to worry about! In fact, wind measurements in the U.S. for over 100 years have shown a remarkable consistency and predictability as discussed in the previous chapter.

Finally, because the lifecycle of modern utility-scale wind turbines and hydrogen electrolyzers is 50 years with minimal maintenance requirements, once it is built, *the cost to produce hydrogen will never change for 50 years.* In fact, as

---

[114] National Geographic magazine, September 2004.

technology improves and becomes less expensive, as it always does, the cost to produce renewable hydrogen will actually drop over time. The cost of generating a kilowatt hour of electricity from wind energy has dropped by 90 percent since 1980 and continues to drop further every year. The same goes for electrolyzers. Therefore, because the energy industry is the largest industry in the world by far and impacts all other industries so significantly, converting to renewable hydrogen can actually become the most powerful hedge *against* inflation that our great nation has ever seen. But many vested interests will resist this change as discussed in Myth #7, so it up to us as individual, freedom-loving Americans to be proactive and make it happen on our own, with or without the blessing of the government or energy industry.

Hopefully this Mythbuster section has helped to alleviate any concern – or even outright fear – that you may have had about hydrogen. So why, if these myths have answers so readily available, do they persist? Perhaps they have been exacerbated over the years by the oil and gas industry. Or perhaps there simply hasn't been a reason to really pay attention until now. Either way, it is time the truth about hydrogen is revealed on a mass scale. Only then, with broad public acceptance by the American public, can we hope to induce change.

Renewable Hydrogen Production Example

To demonstrate the economics behind the production of renewable hydrogen, an actual example of the volumes and measurements of the entire process using today's prices and efficiencies is displayed below. We will show the data of renewable hydrogen produced by an electrolyzer, then used on-site by an alkaline fuel cell or hydrogen fueling station. A 2 kilowatt electrolyzer that runs at 85 percent efficiency uses 600 kWh per month (0.6 MWh) operating 10 hours per day during off-peak hours. It will consume 100 liters of water per hour to create 10 kilograms (kg) of H2 per hour. A 1 MW fuel cell, operating at 68 percent efficiency (compare that to your car's gasoline engine that is a mere 20 percent efficient, meaning 80 percent of the energy content of gasoline is lost – wasted – in the form of heat and exhaust) will consume 0.85 kg of H2 per hour.

1 kg (2.2 lb) of hydrogen has about the same energy content as 1 U.S. gallon of gasoline, which weighs nearly three times more: 6.2 lbs. The U.S. Department of Energy states that "the hydrogen extracted from a gallon of water could drive a

hydrogen fuel cell vehicle as far as gasoline vehicles travel today on a gallon gasoline."[115]

The Hydrogen Economy is coming – what's in it for you?

It is not a matter of if the hydrogen economy is coming; it is a matter of when. The signs indicate it has already begun. A recent article in **FORTUNE** Magazine by David Stipp entitled The Coming Hydrogen Economy says "as far back as Jules Verne, visionaries have predicted that society will someday be utterly transformed by energy based on hydrogen. It is wondrously clean, emitting mainly pristine steam when burned. When fed into fuel cells, which generate electricity, it offers unprecedented efficiency – these electrochemical reactors extract twice as much useful energy from fuel as internal-combustion engines can. *In fact, hydrogen-powered fuel cells promise to solve just about every energy problem on the horizon.* In homes and offices, fuel cells would keep the lights on when the grid cannot. Cars propelled by the cells wouldn't foul the air. None of this is as pie-in-the-sky as it sounds. Potent commercial forces are bringing the hydrogen economy along faster than anyone thought possible only a few years ago. In the next two years, the first wave of products based on hydrogen-powered fuel cells is expected to hit the market, including cars and buses powered by fuel cells, and compact electric generators for commercial buildings and houses. Technology for generating hydrogen is ready now: "reformers" that extract hydrogen from natural gas, and "electrolyzers," Jules Vernian devices that extract hydrogen from plain water. Those electrolyzers, if powered by so-called renewable-energy technologies like wind turbines and solar panels, could truly put an end to oil. Wind turbines and solar panels are emerging fast; after long decades of development, they have entered a Moore's law-like pattern of rapidly falling costs. All these advances add up to a startling reality. Major oil companies have begun to bet quietly but heavily on a hydrogen future. So have many of the largest manufacturers, including United Technologies, General Electric, Du Pont – and every major car company."

According to the Rocky Mountain Institute, "the oft-described technical obstacles to a hydrogen economy – storage, safety, and the cost of the hydrogen and its distribution infrastructure – have already been sufficiently resolved to support rapid deployment starting now. No technological breakthroughs are needed, although many will probably continue to occur."

---

[115] U.S. Department of Energy, Energy Efficiency and Renewable Energy Department "Hydrogen and Fuel Cells Guide"

Surprisingly, hydrogen fuel cells in buildings and businesses (as well as some homes) can today economically displace centralized power plants that deliver power over long-distance power lines. Many large metropolitan areas in the United States have adequate wind resources nearby that could be used to electrolyze water into hydrogen with negligible distribution costs. Finally, early adopters of home hydrogen systems, existing and future state and local government incentives for hydrogen systems, and clever marketing companies will combine to accelerate the installation of numerous fuel cell systems in a distributed manner. Some of these systems will use existing excess capacity in the natural gas and electricity distribution systems to produce hydrogen from natural gas, methane, and propane, while others will utilize small-scale renewable energy. The businesses that identify these niche opportunities to utilize the performance benefits of hydrogen will quickly differentiate themselves from their competitors in both public relations and profitability.

Notwithstanding the progress made by The Freedom Plan, other factors will inevitably feed the hydrogen transition. As long as fossil fuels remain costly relative to the cheap energy prices we enjoyed in the 1990s, and while price volatility persists, the market forces of our democratic and capitalistic society will naturally drive the expansion of the hydrogen infrastructure. And that expansion will be wildly successful; it has to be, if we hope to retain our global position of wealth and strength. A heartening prediction was recently made by the former head of Accenture's $2 Billion global energy practice: a $280 Billion [one-time] investment in hydrogen infrastructure in the U.S. could displace $200 billion per year in annual oil imports by 2020[116]. But this investment, to be effective, should originate from the private sector. Government intervention will simply slow the process down and drastically increase its overall cost.

---

[116] M. Tolan, "Innovative Visions — Art of the Possible: Potential for Dramatic Energy Mix Shift," Accenture International Utilities and Energy Conference (Aventura, Florida, 24 March 2003); R. King, "Mary Tolan's Modest Proposal," *Business 2.0*, June 2003, pp. 116–122, http://www.business2.com/articles/mag/0,1640,49464,00.html.

What will Hydrogen cost if the Freedom Plan is implemented?

The U.S. Department of Energy says bulk hydrogen made and consumed onsite costs about $0.71 per kg,[117] which is equivalent in energy content to gasoline at a mere $0.72 per gallon. But because hydrogen fuel cells are so much more efficient than internal combustion engines, one kg of H_2 used in a fuel-cell vehicle is actually equivalent to gasoline at only $0.36 per gallon![118] This price, however, does not include the cost of delivering the hydrogen to the fueling station. Estimates by the U.S. Department of Energy, various private and government studies in Germany (where BMW is leading the effort to develop hydrogen fueling stations), and many other sources studied indicate that with hydrogen produced at night using utility-scale, wind-derived electricity and delivered to sites no further away than 200 kilometers (about 120 miles), the total delivered cost should be no more than $1.50 per kg of H_2. This equates to about $1.53 per gallon of gasoline in terms of energy content, and $0.75 per gas-gallon-equivalent per mile driven when using a fuel cell.

Imagine…

…being able to buy 75 cent gas,

…that not a single drop of your fuel came from a foreign source,

…that your fuel was *produced* with zero pollution,

…that the fuel you are using *as you drive* creates no pollution,

…using a fuel that actually cleans up existing pollutants in the air,

…a fuel that has a fixed – and stable – cost for 50 years.

---

[117] U.S. Dept of Energy's Hydrogen Information Network, available online at: (www.eere.energy.gov/hydrogenand fuelcells/hydrogen/faqs.html).

[118] Twenty Hydrogen Myths by Amory B. Lovins, Rocky Mountain Institute, 9/2/03 p.6

All the technology is available today. The estimate above is based on today's prices for all components, including utility-scale wind energy facilities with mean capacity factors of 38 percent or better and an installed cost of $1 Million per megawatt. All we need to make this a reality is to implement The Freedom Plan (described in the next chapter). The most powerful method to guarantee implementation of The Freedom Plan is using market-driven forces; relying on politicians to sustain a long-term (10-year plus) commitment to radical change that will undermine the interests of very large political donors is a certain recipe for failure. Of all market forces, consumer demand is by far the most powerful. Once America has enough customers choosing clean power for their homes in the form of Green Tags (see later chapters) and enough customers on waiting lists for hydrogen-powered vehicles (both described in later chapters), The Freedom Plan *cannot be stopped* – by anyone or any thing.

Chapter 6:

THE FREEDOM PLAN

CONVERTING AMERICA TO CLEAN, HEALTHY, RENEWABLE POWER

Once the enormous potential of renewable energy sinks in, some questions inevitably come to mind: is it possible to convert the whole country to clean power? Can we really produce enough hydrogen to power all our cars and trucks? Is it cost-effective? Will it increase energy prices or lower them? How long would it take?

This chapter answers those questions with a realistic plan to convert our nation to 100 percent clean, renewable, American-made power. Supporting facts and financial data will be provided, along with detailed assumptions used in the calculations. It is my hope that by the end of this chapter, readers will have a sense of urgency to do their part to help implement this or any other viable plan that accomplishes the same objective. The bottom line is, we have the technology and the resources to quickly end our reliance on foreign and fossil fuels – the only remaining question is whether we have the collective willpower to make it happen.

FORTUNE Magazine boldly tacked the issue with a Cover Story on August 11, 2004: "How to Kick the Oil Habit." FORTUNE's plan consists of four key approaches:

Improving Fuel Economy.
More Spending on Alternative Fuels.
Redoubled Commitment to Efficiency.
Getting Serious about Wind and Solar.

In the article, it shows how an investment of less than $4 billion per year into hydrogen and ethanol could reduce oil usage by 20%. As you will learn in this chapter, all four of these elements recommended by the FORTUNE plan are already included in The Freedom Plan.

The most significant aspect of the FORTUNE story in my opinion is the underlying theme stated by senior writer Nick Varchaver: "some companies are already taking innovative leadership roles in these areas and it would be ideal to **rely solely on market mechanisms.**" In other words, we cannot rely on any government administration to make this happen. The Freedom Plan is 100 percent market-driven. Although any of the suggested government policies could certainly accelerate the implementation of this plan, it does not rely on the government to be successful.

The Freedom Plan could be completed in 10 years, and consists of five steps:

THE FREEDOM PLAN

1. **Wind Energy**: Add huge amounts of wind power generation in the Midwest and upgrade the transmission grid. This requires new and upgraded power lines and substations, using composite transmission lines with fiber optic cables embedded into the lines, allowing sophisticated real-time 'neural network' monitoring of the grid.

2. **Renewable Hydrogen**: Install utility-scale electrolyzers (to cleanly produce hydrogen from water using wind power), hydrogen storage tanks, and alkaline fuel cells at strategic locations across the national grid. Many of these locations will be in close proximity to depleted oil and natural gas fields where huge quantities of hydrogen can be stored cost-effectively. Begin pumping hydrogen into existing natural gas pipelines (initially blended with natural gas into Hythane, and eventually as pure hydrogen). Construct new transcontinental hydrogen pipelines along the rights of way of transmission lines, as well as new regional and local pipelines where needed. Add sources of renewable methane to the natural gas system.

3. **Energy Efficiency and Distributed Generation**. Launch a massive distributed generation and energy efficiency campaign, where homes, businesses, industry and government buildings install solar generation, fuel cells, small wind projects, renewable methane, ocean energy technologies and other renewable power systems. Cut the overall energy consumption in the United States by at least 30 percent in three years, 50 percent in ten years.

4.   **Clean up Existing Power Plants**. Convert all remaining power plants to clean, renewable fuels. Modern coal plants should burn biomass and natural gas plants should burn a) renewably-produced hydrogen and/or b) renewably produced methane (landfill gas, municipal wastewater, industrial waste, and livestock facilities). Older inefficient power plants should be retired or upgraded to clean renewable generation and then used exclusively for summer peak power generation and backup power.

5.   **Hydrogen Fueling Infrastructure**. Add compressed natural gas and compressed hydrogen fueling pumps at gas stations nationwide. As this fueling infrastructure is built out city by city, the market will naturally be created for people to convert their existing vehicles to burn hydrogen by installing conversion kits. Furthermore, as hydrogen delivery becomes available, other related products will become widespread and popular: fuel cells that will provide safe and reliable 'in-home power plants' for homes, commercial fuel cells, and residential hydrogen production systems (using solar shingles and other interesting products).

This 5-step Freedom Plan addresses the conversion in two primary phases: 1) convert the generation of electricity (through the grid) to clean power and then 2) convert the transportation sector to clean, renewable fuels. There is some overlap between these two phases, but it would be nearly impossible to provide clean renewable fuels for vehicles without first converting electric generating plants to produce renewable power and building renewable Hydrogen (H2) production facilities with the related H2 infrastructure. We cannot power our vehicles with electricity from the grid without creating enormous amounts of new pollution (remember, power plants belch far more pollution into the air and water than every car and truck in America today). Therefore, we must clean up the electric generation system using a method that will automatically lead to a clean, renewably produced transportation fuel. The only known mechanism to accomplish this objective is hydrogen. And the only sensible way to produce hydrogen is using clean, renewable energy from cost effective sources. Today, the most cost effective source is wind. After a few years of building out The Freedom Plan, other power generation forms such as solar, ocean thermal and/or tidal generation or other more elegant sources of energy may become more inexpensive than wind power. But to 'kick-start' the plan and gain the critical momentum necessary for success, wind energy is the only source commercially available today that has a cost of energy that is adequate for the success of The Freedom Plan.

As discussed in Chapter 3, our total electricity demand in the United States is roughly 4,000 terawatt hours per year. Because the creation of this electricity is the largest single source of pollution in this country – greater even than vehicle emissions – it is imperative that we first provide a clean source of electricity. Powering 100 percent of the electric grid with renewable power is a huge undertaking. Many of my critics will emphatically say it is not possible. Not only is it possible, it is necessary. Due to the daunting challenges, this conversion is going to take time, a great deal of effort, and consistent ongoing support from the public. Here is the Freedom Plan.

STEP 1

Phase 1

Begin creating a public groundswell of support and acquire at least five million homes paying a little extra for their power in the form of Green Tags. As of 2004, about one million Americans are paying more for their power to make it clean and renewable. In Europe, there are nearly 50 million customers of clean power who pay more for the energy by some measures. To stimulate enough development of wind energy, The Freedom Plan calls for five million Americans buying Green Tags by paying an extra ten to thirty dollars per month for their power. Why will people do this? While it may seem unlikely that individuals will pay a premium for clean energy, people will do so because it is the right thing to do and because it is the only way to guarantee the conversion to clean, renewable power. This will stimulate the renewable energy industry and inject the necessary capital to successfully launch this campaign. Subsequent chapters will describe in detail how you can help usher in the Renewable Age.

Phase 2

Assuming a starting point of 100,000 MW (megawatts) of installed or planned wind energy capacity, The Freedom Plan adds a whopping 2,000,000 MW of new wind energy capacity over **10 years** (see Table 2). Today's cost for installed wind energy facilities is roughly $1 Million per MW with an average delivered energy price of 4 ¢ per KWh (kilowatt hour), including many smaller wind projects, which are not as cost-effective. These costs are dropping consistently. A large project in a great wind area can deliver power at 3 ¢ per kWh (or as low as 1.5 ¢ per kWh with the federal Production Tax Credit). In fact, in 2004 General Electric (the wind turbine manufacturer with the largest market share in North America) set a short-term goal of rapidly reducing the energy price by one cent per kWh.

A good year in the U.S. wind power business has been the addition of a meager 1,500 MW. Therefore, adding two million megawatts in ten years sounds like an enormous increase. Make no mistake, it most certainly is. This requires a massive commitment to wind energy, focused in areas where wind power is embraced (primarily the Midwest). With a consistent, assured market of hundreds of thousands of MW of new wind energy equipment sales every year, wind turbine manufacturers will make major capital investments in manufacturing, design, and component technology. This will significantly reduce costs even further, making wind energy even more competitive than it already is. Combine those cost savings with the economies of scale of such huge additions of wind energy and the cost should easily come down to $600,000 per MW. In fact, the cost would probably be less than $400,000 per MW, but in the interest of keeping a conservative theme I have chosen to prove how this conversion is economic even with higher cost assumptions. Thus, at an average cost of $600,000 per MW, adding two million MW of wind energy will cost **$1.2 Trillion** or 60 percent of the total cost of The Freedom Plan.

How much electricity will these wind farms produce? In good wind areas of the Midwest today, one can expect a capacity factor of 35-45 percent (this means a 100 MW wind farm would produce 100 MW at times, 0 MW at times, but 35-45 MW on average throughout the year). Capacity factors could reach as high as 45 – 51 percent in the best areas. The world record for capacity factor is 51.7 percent at a wind farm in New Zealand. In other areas of the world, capacity factors of 30-35 percent are to be expected with today's larger more efficient turbines. Building these projects in the best wind areas of the Midwest, and assuming at least incremental improvements in technology to increase net capacity factors, it is safe to assume an average capacity factor of 40 percent. Again, this is likely a conservative estimate since there are hundreds of thousands of acres in the Midwest that are already leased for wind farms that would have average net capacity factors of 43 percent or higher. Also, most of the best wind sites are in areas where the local farmers and ranchers are eager to see new wind farms due to the economic impact they receive. Adding two million MW will not require a huge amount of land, either. A recent study by Garrad Hassan found that every home in Europe (150 million homes, compared to only 100 million homes in the U.S.) could be powered by an off-shore wind farm covering an area the size of Greece, a small country in the Mediterranean. To power all of the United States with wind, it would only require an area covering about one-half of a typical Midwestern state. In reality, however, these wind parks will be spread out over dozens of states.

An average site should generate royalties to the landowner of $2,000 to $5,000 per year per wind turbine. A rancher might make $10-25 per acre grazing livestock and a farmer may generate $15-55 per acre growing crops. Hosting wind turbines can add another $75-120 per acre, plus the landowner can still graze cattle or farm the land right up to the base of the wind turbine. Therefore, with hundreds of thousands of acres of excellent wind resource available, there are plenty of areas where little if any resistance to building these projects will be encountered.

With two million MW of wind generation facilities operating at a 40 percent capacity factor, a total of 7,358 terawatt hours will be produced annually from 100 percent clean, renewable, American-made wind power. This represents 92 percent of the goal of 8,000 TWh per year of total energy. Since part of the plan includes a significant reduction in total energy demand and an ample supply of distributed generation (which reduces the demand for power from the grid significantly), this 92 percent figure actually allows for an increase in power demand in the first few years of The Freedom Plan. From 1998 to 2003, average electricity consumption has increased at about 2 percent per year. However, electricity conservation and energy efficiency programs (Step Three) will actually reduce total demand significantly (as seen in California in 2001-2003 when total electricity demand fell by over 14 percent in just 2001 alone[119]). Today, America consumes the equivalent of about 10,000 TWh per year of total energy (4,000 TWh of electricity, 6,000 TWh of fuel). The Freedom Plan will cut this demand to about half of that total, or 5,000 TWh per year with energy efficiency, distributed generation, and lighter more fuel efficient hydrogen-powered vehicles. But the plan includes total generation capacity of over 7,000 TWh, well over what is really necessary. This is by design. The extra capacity allows for the maximization of wind energy generation, producing power for the grid when it is needed and hydrogen when demand is lower. The hydrogen can be stored to increase energy security, add energy storage capabilities (which improves grid reliability and efficiency), and stockpile large quantities of clean transportation fuel. But the most significant element of this extra generation capacity is the ability for America to begin exporting energy in the form of hydrogen. America can – and should – become a net energy exporter for the first time in over three decades. We can utilize our powerful native winds to create clean hydrogen and help the rest of the world convert their energy infrastructure to clean power using a similar model to The Freedom Plan. The importance of the positive economic impact of such a huge export market cannot be overstated. This could easily reduce or trade deficit by hundreds of billions of dollars per year.

---

[119] California Flex Your Power program, at http://www.fypower.org/about/faq.html.

Actually, it could perhaps be the only solution large enough to actually create a trade surplus.

Phase 3 – OK, now comes the hard part. Building the wind energy facilities is easy compared to the challenge of connecting all that additional generation to the national transmission grid and moving the power to the coasts. Our electric grid is weak, outdated and in desperate need of upgrade. It was built primarily using technology developed in 1898 and most of it was constructed in the 1950s and 60s. Because wind generation was not used when the transmission gird was constructed, the best wind areas, which are in the rural Midwest where population is sparse, do not currently have enough high voltage transmission lines installed. Therefore, to enable the construction of massive amounts of new wind energy generation, new, high-capacity transmission lines are mandatory. Furthermore, upgrading existing power lines and substations is already a necessity and will soon become even more critical.

According to a TIME magazine article[120] after the 2003 blackout in the northeast, "the best way to upgrade the energy grid may well involve doing away with some of it, democratizing energy production by handing the job off to communities, blocks and even private homes...in the form of 'power parks' – communities or mere groups of homes that would generate their own energy courtesy of solar panels, wind turbines, fuel cells or natural gas generators. The little clusters could be almost entirely self-sufficient, relying on the grid only in the event that they needed to top themselves off with a sip or two of outside power. Just as important, they would have the freedom to disconnect from the large network entirely if a regional crash was threatening to knock them off-line along with the bigger consumers. Similar systems could supply factories or hospitals."

The Freedom Plan calls for building 170 new high capacity 750 kV (kilovolt) transmission lines at a cost of almost $200 Billion (10 percent of the total of $2 Trillion). Simply upgrading existing transmission and building several new local transmission lines to remove system bottlenecks and improve system reliability to add 26,000 MW of new wind energy would cost a mere $1 Billion.[121] This $1 Billion cost is enough wind energy to meet the requirements for nearly all of the first two years of new wind installations as outlined in The Freedom Plan. It is also enough to reduce the amount of natural gas our nation uses by 2.4 Bcf/day[122] (2.4 billion cubic feet per day), or a reduction of about 10 percent of current

---

[120] TIME Magazine, August 25, 2003, pp. 36-37.
[121] AWEA "Wind Power Outlook 2004" p. 5.
[122] AWEA "Wind Power Outlook 2004" p. 5.

natural gas consumption. And, it is enough wind energy to power 6.5 million average American homes.

That initial $1 Billion investment would be a great start, but the next step would be to build two major high voltage lines from the Great Plains of the Midwest to both the east and west coasts (four lines total) at a cost of about $15 Billion according to a study performed for the American Wind Energy Association. These initial steps could be expanded upon by adding additional lines to the existing structures built for these power lines (assuming the design was performed to allow for such expansion) at a greatly reduced incremental cost. The Freedom Plan's 170 new power lines, many of them much shorter than the primary lines going to both coasts, can be built along existing power line rights of way to minimize permitting costs and approval hurdles. And many existing lines can be reconductored, which means taking the old line off and replacing it with a new, higher capacity composite line.

NEW POWER LINES

- 30 lines from Colorado to Ohio at $700 Million each
- 20 lines from North Dakota to Texas at $1 Billion each
- 20 lines from western KS/OK to Chicago at $1.5 Billion each
- 100 lines elsewhere at $1.25 Billion each
- Total: 170 new lines at a total cost of **$196 Billion**

These new power lines, constructed during the first **6 years** of the Freedom Plan (see Table 1), will connect the eastern and western grids together, as well as the Texas (ERCOT) grid. The lines should use the latest composite technology and have a fiber optic channel integrated into the system for advanced monitoring and operational control, acting as a neural network power system. This type of system would instantly detect if a span goes down, and reroute power instantly to other lines, protecting itself from blackout risks. The blackout of August 2003 in northeastern North America cost the United States between $6.8 and $10.3 Billion, according to estimates by ICF Consulting.[123]

It is critical to improve the 'stiffness' of the grid to accommodate wind power in massive doses. This means having sufficient transmission capacity to spread the wind variability – and the load variability – over a wide region. It also includes the ability to scale up or scale down other forms of generation based on signals from wind forecasting software, spot electricity price maximization systems, and grid operators.

---

[23] "The Economic Cost of the Blackout (an issue paper)" by ICF Consulting, Fairfax, VA.

The cost for these power lines is overestimated on purpose to allow for inflated costs for rights of way and permitting. For example, a new 750 kV line from Denver to St. Louis is estimated cost about $500 million today. Economies of scale of 170 new lines would lower that cost significantly – the actual cost of the Denver/St. Louis lines would probably be less than $400 Million, almost 50 percent lower than the $700 Million included in the conservative Freedom Plan. Also, many states, including Kansas in the center of the country (and the top-ranked state for renewable energy potential), have eased the permitting requirements for new transmission lines, lowering the cost of acquiring rights of way and associated legal expenses.

In addition to these new lines, upgrades to existing power lines and substations are necessary. In fact, many existing lines could have their capacity doubled or tripled by simply replacing the obsolete wire with newer composite cables with much greater capacity – reusing the same poles and structures as indicated earlier. Larger transformers and additional switching equipment will be necessary at the many substations along the way. These upgrades have to occur anyway, so only a portion of those costs are included in this Freedom Plan. However, since there is plenty of 'wiggle room' in the $196 Billion cost for new power lines, all of these upgrades could likely be included as a part of the Freedom Plan.

STEP 2

Phase 1

The only real challenge with wind power, from the perspective of managing the power grid, is its intermittent nature. The wind is very predictable, but not reliable (see Chapter 4). Therefore, a cost-effective method to store wind energy is needed. Fortunately, such a method exists that has a dual purpose: energy storage AND a clean fuel that can be produced with renewable wind and solar power.

Hydrogen (see Chapter 5) solves both problems. Hydrogen, like oil or natural gas, is a form of stored energy that can be burned in an internal combustion engine (in vehicles or power plants) or used in a chemical reaction in a fuel cell to create electricity. But unlike oil or natural gas, it can be produced cleanly, using renewable energy, allowing us to store wind energy when it is not needed (during nighttime off-peak hours) for later use during peak hours to 'balance the load' of wind energy facilities.

Of course, injecting hydrogen into our existing natural gas pipeline infrastructure can also help alleviate the current natural gas crisis (which is projected to steadily get worse over the next 20 years). In addition, it can supply a renewable, American-made fuel for the transportation sector. Natural gas powered vehicles already exist (these vehicles are very easy to convert to hydrogen) but are not commonplace because there are not enough compressed natural gas (CNG) fueling stations. Because natural gas is becoming a scarce resource, it does not make sense to switch all our vehicles to CNG. Hydrogen, on the other hand, can be plentiful and inexpensive, as long as it is made at night during off-peak hours from wind farms or other renewable sources. Renewable power is combined with water in a simple process called electrolysis invented in the year 1800 (over 200 years ago) by the English chemist William Nicholson and then enhanced by Polish inventor Johann Wilhelm Ritter. In fact, the water source can even be municipal wastewater, solving another problem.

Large-scale electrolyzers can be purchased today for as little as $1 million per MW. This includes the 'balance of plant' components, such as compressors (hydrogen is such a light gas, to store it in useful containers is must be compressed or liquefied), pumps, storage tanks, filters and purifiers. Companies like Stuart Energy, Avalence, Norsk Hydro and others have steadily growing sales of electrolyzers and estimate that with enough economies of scale from mass production (at least 5,000 MW per year) and more competition, the costs could drop to 15 percent of that figure ($150,000 per MW or $150 per kW).

Locating these electrolyzer plants is another challenge, but one that can easily be overcome. Factors to consider in locating electrolyzer plants are 1) access to inexpensive electricity during off-peak (nighttime) hours; 2) access to inexpensive water; and 3) ability to store the hydrogen in large quantities (ample land for large tanks or other storage methods).

As discussed in Chapter 5, these factors all have cost-effective solutions today. Many of these plants will be located near depleted natural gas fields, where the H2 can be injected back into those fields for large-scale storage. This has the added benefit of squeezing out residual natural gas from the fields that had previously been considered unrecoverable. The largest natural gas field in the world is the Hugoton gas field in southwest Kansas (it is so large that it extends into Colorado and the Oklahoma and Texas panhandles). This gas field is expected to run 'dry' by 2010. Incidentally, this is also one of the windiest places in the world – with wind farm capacity factors expected to average 45-53 percent. This one area of the country could power over 30 percent of the nation's current energy needs with wind and hydrogen.

Once the hydrogen is created and stored, the next challenge is how to use it to create electricity. Internal combustion engines, such as those in vehicles, typically run at 15-25 percent efficiency (this means 75-85 percent of the energy in the fuel is wasted as heat, exhaust, emissions, etc.). Hydrogen powered fuel cells are 50-70 percent efficient. With ample supplies of H2, a market will quickly emerge for fuel cells in homes, businesses, hospitals, schools, and power plants.

Today's cost for alkaline fuel cells (originally invented in 1839 by Sir William Grove, a Welsh judge and gentleman scientist) are about $2 million per MW, with prices expected to drop to $500,000 per MW ($500 per kW or 50¢ per watt) or less by 2005 with current sales projections. PEM and other fuel cell technologies are more expensive than alkaline, but prices are also dropping quickly. Discussions with existing suppliers reveal that nearly all of them have the ability to scale up to mass production in short order. In fact, it is estimated that with orders of just 1,000 MW per year, fuel cells will drop in price to $150,000-400,000 per MW. With mass production, the cost should drop to $50,000 per MW.

Furthermore, it is important to note that because many people will take their homes 'off grid' with solar shingles and small fuel cell appliances, the demand for electricity at power plants will be reduced. This is more efficient since line losses are lower (moving large quantities of energy from centralized power plants over long distances is inefficient). Investment in new distribution power lines can be reduced. Where new power plants are absolutely necessary, large-scale fuel cells can be installed to convert H2 into electricity while older coal and gas fired plants can be taken off line.

New power plants will have to be constructed with or without Tthe Freedom Plan in order to meet increases in demand and to replace aging plants, so including the cost of all the fuel cells necessary to power the grid when the wind is not blowing is not appropriate. As fuel cells are added with distributed generation (in homes and businesses), those costs will be paid by the end user and do not need to be included in the Freedom Plan. However, to be conservative the Freedom Plan includes the cost to add 80 percent of the 2.1 million MW of wind energy, or 1.68 Million MW of electrolyzers and fuel cells (where needed) at an average cost of $300,000 per MW for a total of $504 Billion

(25 percent of the total cost of The Freedom Plan). Incidentally, this could produce 840 million kilograms of hydrogen per day if the electrolyzers were in use fulltime. Readers may remember from Chapter 5 that one kilogram of H2 is roughly equivalent to one gallon of gasoline. In 2003, America consumed exactly 840 million gallons per day of gasoline, diesel, and other petroleum products. In addition, $100 Billion is set aside for construction of new hydrogen pipelines (or for upgrading existing natural gas pipelines where needed) and hydrogen storage. The phase in of this hydrogen infrastructure and equipment would begin in year two of the Freedom Plan and increase steadily over nine years (see Table 1). Existing power plants will not be shut down immediately, so many of them will remain available for baseload power and peak demand shaving. The total H2 production and fuel cell power generation capacity included in The Freedom Plan is more than what is needed to power the entire electric grid and every car, truck, train and plane in America assuming the overall reduction in energy demand of 50 percent created by energy efficiency and distributed generation in Step Three of the plan. This also takes into account the losses inherent in converting wind energy to hydrogen and then back into electricity in the fuel cells, as well as the modest losses of transporting hydrogen around the country in pipelines as it becomes the new primary transportation fuel. The total cost of fuel cells, electrolyzers, storage, and H2 pipelines: **$604 Billion** or 30 percent of the total.

Although many newer cars will be much lighter and more efficient carbon fiber vehicles that will require far less hydrogen fuel, The Freedom Plan calculations are intended to be conservative enough to accommodate a large percentage of current total energy demands. With the added efficiency of lighter fuel cell vehicles and lower overall power demand, The Freedom Plan will have ample overcapacity in the system for growth and large spikes in energy demand.

Since the wind farms will produce an average of 40 percent of their capacity to generate more than enough power for the grid, if at any time there is no wind blowing at any wind farm anywhere in the country (an impossible scenario) there would be enough backup power available. Remember, wind energy can be predicted using sophisticated forecasting models (these are available today) to ascertain the exact output of any wind farm anywhere in the world 24 hours in advance with an accuracy of plus or minus 5 percent. Therefore, we will know in advance when the wind is not going to blow strong enough to meet demand, allowing ample time to notify power plant operators to scale up their generator to meet demand. Note: fuel cells can be online within seconds, and can be programmed to start up electronically without human intervention whenever the wind dies down.

In summary, the Freedom Plan includes more wind generation than we really need, assumes a lower net capacity output than what will be realized, adds more transmission lines than what is really necessary at a higher cost than what is likely, includes electrolyzers for creating more hydrogen than we really need, and adds more fuel cells than what we need to back up the wind power at a cost higher than expected. All for a total price tag of $2 Trillion.

Suggested timeline of wind farm, transmission, and electrolyzer additions:

Table 1

Year	MW of Wind Power	Power Lines	MW of Electrolyzers
1	2,325	5	
2	32,985	10	100
3	40,816	15	500
4	40,740	20	5,000
5	56,825	30	25,000
6	87,314	30	100,000
7	184,727	30	300,000
8	428,523	10	500,000
9	728,449	10	750,000
10	1,138,705	10	0
Total	**2,741,410**	**170**	**1,680,600**

As shown, the transmission line infrastructure, the wind farms and hydrogen equipment are built over the 10-year period of the Freedom Plan. This is a timetable that is realistic and attainable, using technology available today. The total price tag is a big number ($2 Trillion) but pales in comparison to the cost to America if we do not convert to clean power. More information about the savings to our nation will be addressed in the next chapter.

Natural Gas Crisis

As this clean, renewable source of hydrogen is being constructed, an important task of the Freedom Plan is to solve the existing natural gas crisis. Extra hydrogen created (beyond what is needed for backup power and on-site storage at the key fuel cell power stations along the transmission grid) can be injected into the existing natural gas (NG) pipeline system. When blended with NG, the hydrogen/NG mixture is called Hythane. Since NG is over 90 percent methane, and methane (CH_4) is comprised mostly of hydrogen, in most cases no modification is needed on appliances and engines that currently use NG (assuming the Hythane blend is no greater than 20 percent hydrogen).

There are a variety of ways to produce renewable NG or methane. For example, agricultural facilities (cattle feedlots, hog farms, chicken farms, etc.) produce a great deal of methane. Landfills all over the world are a great source of consistent methane, which can be collected using methane gas wellheads. Municipal (city or county) wastewater produces a great deal of methane. Finally, some industrial processes produce waste methane. A few of these sites collect this methane presently, using anaerobic digesters, gas wellheads and other techniques.

With better education of landfill and wastewater operators, industry, and rural livestock managers, the implementation of these methane collection technologies will become more widespread. As more are installed, the costs will continue to decline. There are even devices available for homeowners that act like a miniature anaerobic digester to collect methane from food waste to inject it back into your home NG system. It is estimated that using the techniques mentioned above to collect only the most cost-effective methane, more than 30 percent of America's current NG needs would be provided for.

Because our nation relies so heavily on NG, the infrastructure is already built to move NG around the country. New NG pipelines are being constructed today with even more being planned to meet demand. Unfortunately, our domestic supply of NG is in danger. Therefore, the next step is to begin pumping hydrogen through existing NG pipelines. The NG grid is extensive, and little modification is necessary to move hydrogen through this system. In fact, there are standard steel tanks (300 psi) built in 1907 that are still holding hydrogen today without any leaks or embrittlement. See Chapter 5 for more information about hydrogen.

By far the biggest challenge of using hydrogen as a primary source of energy storage and fuel for vehicles is the transportation and delivery of hydrogen. Because H2 can easily be pumped through most existing NG pipes or blended with NG into Hythane, the mechanics of this step three are relatively simple and the costs are negligible. The real challenge will be overcoming the resistance of the fossil fuel mentality that permeates the gas pipeline industry and educating pipeline operators about the facts (dispelling the myths) of hydrogen.

It should take no longer than 5-7 years to be a) producing large quantities of renewable methane that is injected back into the NG infrastructure; b) injecting H2 into NG to create Hythane and utilize this gas in nearly all existing NG appliances and applications; and c) transform some existing pipelines to 100 percent hydrogen once enough H2 is generated using wind-powered electrolyzers. See the Timeline for a suggested realistic conversion timeline.

As the need for more hydrogen increases during Steps Four and Five, additional hydrogen pipelines can be constructed – if necessary – along existing rights of way where NG pipelines and transmission lines are constructed. As current NG users begin switching to all electric homes (in order to power their homes more efficiently with fuel cells), the demand for NG will decrease, freeing space for more H2 in the existing pipelines.

STEP 3

ENERGY EFFICIENCY

Along with converting our electric generation capacity to wind and other renewable sources, the total demand for electricity must be addressed. In 2001, the state of California launched a conservation and energy efficiency educational campaign. It was wildly successful, lowering peak electric power demand by nearly 14 percent in just one year in the state with the largest economy in America, by far. Energy efficiency is easily the quickest, cheapest, and cleanest source of new energy.

A recent BusinessWeek editorial said "we need a new plan to reduce U.S. dependence on oil from overseas."[124] You are obviously reading about such a plan. Perhaps someone should tell the editorial staffs at these various business publications about it. The BusinessWeek editorial also stated that we must "encourage conservation...hybrid cars...and fuel-cell-powered cars" and that we should use "solar power [wind is a form of solar energy as discussed in Chapter 4] and other alternative sources...to generate the hydrogen."

Energy efficiency is not conservation. Asking people to turn down the heat and put on a sweater is an immediate response to the crisis, but it cannot be sustained in the mid-to-long-term. And that requires a lifestyle change, which is much more challenging than simple efficiency techniques. Modern technologies can reduce energy use, without reducing services, by 50-75 percent over a 10-year period by some estimates. The American Council for an Energy Efficient Economy (ACEEE) recently completed a study that shows a 24 percent reduction of all electricity usage could be realistically achieved in the short term in the U.S.[125] This represents a reduction of 100,000 megawatts of baseload generation capacity, or a 250,000 megawatt reduction of required wind energy capacity. However, as we have learned earlier in this chapter, The Freedom Plan does not call for reducing the planned 2 million MW of wind energy capacity, but instead utilizes the overcapacity to produce hydrogen. Any progress in energy efficiency will simply accelerate the completion of The Freedom Plan. Investing in new technologies pays off quickly – especially with today's high prices. ACEEE also said efficiency programs could result in a 20 percent reduction in natural gas use is possible in the short-term, and would inject $100 billion back into the U.S. economy.[126]

California has been a world leader in energy efficiency, cutting demand growth to half that of the US. Over the last 20 years, California's energy efficiency programs have created historic savings - 10,000 megawatts (MW) of power - one-fifth of California's peak demand, and the equivalent of 20 large power plants. One-third of the state's commercial customers and 33 percent of residents cut energy use by at least 20 percent.[127] California's demand increased only 13 percent since 1988, half the national average. California's per capita energy use fell to the lowest of all industrialized nations. If California were a nation, it would have surpassed Japan's impressive energy efficiency record in 2001.

---

[124] BusinessWeek, "Oil: What Must be Done" May 17, 2004, p. 144.
[125] "The Technical, Economic and Achievable Potential for Energy Efficiency in the US: A Meta-Analysis of Recent Studies" by ACEEE, September 2004.
[126] Testimony of William R. Prindle to Joint Economic Committee, October 7, 2004.
[27] California Flex Your Power program, at http://www.fypower.org/about/faq.html.

Energy efficiency can be the cheapest power option. A 1999 Massachusetts energy efficiency program invested $159 million in energy efficiency programs in 1999 and calculated that the cost of saved electricity was 4.2¢/kWh. This can be compared with their average retail price of electricity of 10.2¢/kWh. Clearly, energy efficiency has a compelling return on investment, and is a large part of the success of The Freedom Plan.

Many homes, businesses, schools, churches and buildings will install hydrogen-powered fuel cells to provide clean on-site power generation (this is called distributed generation). Heat and a source of pure drinking water are additional by-products of this process. The hydrogen fuel for these fuel cells will be provided from one of at least three sources: 1) reforming hydrogen out of the existing natural gas/renewable methane pipeline infrastructure; 2) new hydrogen delivery systems (H2-only pipelines in neighborhoods and communities, on-site storage tanks refilled monthly by H2 delivery vehicles, etc.); and 3) hydrogen produced renewably on site.

As these on-site fuel cells become more widespread, the need for centralized electric power plants will be dramatically reduced. And, the power line losses inherent in our current centralized system will be reduced, making our power system more efficient and creating much less waste. In fact, the author has estimated that a 5% reduction in electric demand by using residential (distributed) renewable hydrogen production should lead to a 15-20% reduction of electric generation at centralized power plants due to the benefit of the multiplier effect of distributed generation versus inefficient centralized power production. This will decrease the demand for electricity, but will incrementally increase the demand for hydrogen, further assisting the acceleration of the hydrogen economy.

RESIDENTIAL HYDROGEN PRODUCTION

On-site residential H2 production is not yet affordable today for most people, but with mass production it could be. A recommended method works like this:

1. Collect rainwater from your existing gutters using a rain barrel system (the barrel itself can be located out of site in your basement or garage). City water or well water can be used as a back up. Expected life: 50 years. Approximate cost: $500.

2. Add a small residential wind and/or solar system (3-10 kW in size, with an output of 24 volts DC) to create renewable electricity. This costs today $8,000-$15,000 for a 2 kW wind system, $30,000-$60,000 for a 5 kW solar

system. Because most areas are not windy enough (and zoning limits the use of even small wind turbines), solar PV (photovoltaic) shingles or PV panels are the only practical solution in most cases. Expected life: <u>20-30 years</u>. Approximate cost today for a 5 kW solar PV array: <u>$50,000</u>.

3. Add an electrolyzer to electrolyze the rainwater using wind or solar power during 'off-peak' hours to create at least 1-3 kg of hydrogen per day. Preferably, the output of the electrolyzer would be at a high pressure to allow for compressed H2 gas storage on site. Expected life: <u>50 years</u>. Approximate cost today: <u>$45,000</u>.

4. Add a hydrogen storage tank that holds at least 10 kilograms of H2 on site (enough to run your fuel cell for several days without producing any additional hydrogen). Expected life: <u>100 years</u>. Approximate cost of a stationary 2,500 psi H2 tank: today <u>$8,000</u>.

5. Add a 2-10 kW fuel cell to convert hydrogen to electricity (the process will also create pure drinking water and some waste heat, which can be captured to heat your air or water in your home). Expected life: <u>10 years</u> (with replacement of stack membrane once every 3-5 years at an additional cost of $1,500-$2,000 each time). Approximate cost today for a 5 kW fuel cell: <u>$25,000</u>.

6. <u>Optional</u>: add a residential hydrogen-fueling compressor in your garage to fill up your vehicle with compressed H2. For example, the Phill device, designed for compressed natural gas refueling from home, takes about 4-5 hours to fill up a vehicle at night. Expected life: <u>20 years</u>. Approximate cost today: <u>$3,000</u>.

7. <u>Optional</u>: convert your car to burn hydrogen (see Chapter 5) instead of gasoline, diesel or natural gas (NG). Converting a natural gas powered vehicle is much less expensive than converting a typical ICE (internal combustion engine). Approximate cost today to convert one ICE vehicle: <u>$25,000</u>. Cost to convert 50 ICE vehicles: <u>$10,000 each</u>. Cost to convert 1,000 ICE vehicles: <u>$6,000 each</u>. Cost to convert one NG powered vehicle today: <u>$5,000</u>. Cost to convert 10 or more NG vehicles today: <u>$2,000 each</u>. 1,000 CNG vehicles, less than <u>$1,000 each</u>.

If you include the design and installation expenses, shipping costs, and all necessary extra parts and components (excluding the two optional components for vehicles), the total system costs over <u>$200,000</u> today. One company offers such a turnkey home hydrogen system for $150,000 with financing options. Some people find this price acceptable today, especially in off-grid (remote) applications, such as mountain cabins and/or island resorts. Purchases of such systems help to support this burgeoning industry. [For a list of companies and individuals offering such systems for homes and vehicles, contact the author]

However, with enough homeowners willing to join a waiting list for such a system (the author has calculated that it will take approximately 5,000 homes) the cost should drop to $25,000 (less than the cost of a mid-range new car) or less for a complete turnkey residential system (including the H2 fueling dispenser in your garage).

However, before a home could realistically be fitted for a home hydrogen system, the home must be as energy efficient as possible. Today there are enough interesting products and unique technology to cut the average energy consumption (and associated utility bills) by up to 40% on new homes and up to 70% on older homes. There are a number of energy efficiency experts operating businesses in many regions of the country [contact the author for a list in your area] who can provide an inexpensive analysis of what you can do to your home to save money and energy. Some examples of effective products are:

- Tankless hot water heaters: for as little as $350 you can eliminate your hot water tank (and sell it to someone else if you wish) by replacing it with a small 'Instant On' tankless hot water heater that flash heats the water only when you need it. A hot water tank must continue working to heat water while you are sleeping, at work, and on vacation when you don't even need hot water. Heating water on demand can cut water heating costs by 60%, saving you 12-20% (if you use an electric hot water tank) or 8-15% (for gas water heating) at today's energy prices. As energy costs continue to rise, these systems (which often have a 15-year warranty) will save you thousands of dollars over their life., and over pay for themselves in about one year.

- Geothermal ground source heat pump: to heat and cool your home, this device uses the constant temperature of the earth (around 55° F) to supplement or replace your forced air system. These units operate at heating efficiencies 50 to 70% higher than other heating systems and cooling efficiencies 20 to 40% higher than available air conditioners. That directly translates into savings for you on your utility bills.

- Passive solar: adding tile or other heat absorbing materials to the floor of rooms with southern exposures can absorb free solar energy to help heat your home is one example of passive solar.

- Shower heads: there are modern shower nozzles that are used in luxury spas that inject air bubbles into the water stream, increasing the strength of the water jet (it feels like a lot of water) but actually uses far less water. This $15 product cuts down on hot water usage significantly.

- Programmable Thermostats: installing a $100 quality programmable thermostat allows you to set the heating and cooling needs for your family based on time of day and day of week. For example, if no one is home during the day, why heat the house in the winter or cool it in the summer as if you were there? This simple device can begin adjusting the temperature 30 minutes before your family begins returning home from work and school on weekdays, and again prior to waking in the morning to keep your furnace and A/C unit working only when needed. This one product can save you up to 30% on your energy bills alone!

- EnergySTAR® appliances: when it is time to replace an appliance in your home, look for energy efficient models that can save you significantly more in energy costs than standard less expensive models.

- Better windows: obviously, better insulation and better windows and doors keep the warm air in or out, depending on the season.

- Appliance savers: there are simple devices for around $30 that can plug into the outlet where your appliances are (especially refrigerators) that can reduce power demand by up to 40% on that appliance depending on its age. Appliances use the bulk of the electricity in most homes.

- Fuel saving devices: fuel catalysts exist that are not liquid additives that really work and have been tested and certified by numerous independent labs. Although they are hard to find, I have a product on all the vehicles in our household that improves our fuel economy by an average of 20%. This saves our household nearly $85 per month (with gas at $2.25 per gallon).

- Lighting: finally, it amazes me that most Americans still use the same type of light bulb as the one that was invented by Thomas Edison in 1879. Modern technology has developed a new light bulb, called a compact fluorescent light bulb (CFL), that burns with only 20% of the power while providing the same amount of light, emits almost no heat (heat from light bulbs is a big drain on air conditioning systems in the summer), and lasts up to 15 times longer than a comparable incandescent bulb (meaning you rarely have to change them). Each year in the United States, we throw away 1.741 BILLION incandescent bulbs.[128] If every family in the U.S. replaced one regular light bulb with a CFL, we'd reduce global warming pollution by more than 90 billion pounds, the same as taking 7.5 million cars off the road.[129] So replace your incandescent light bulbs with more efficient compact fluorescent lights, which now come in all shapes and sizes.

Once a home is energy efficient, it makes sense to take things to the next level and begin replacing the need for grid-supplied power with on-site renewable energy generation and a home hydrogen system. Exciting products are now affordable such as solar shingles (that look like composite shingles but have built-in flexible solar PV thin-film cells) and roof-mountable small wind generators. Wind systems can pay for themselves in 3-8 years depending on your wind resource and government incentives, while solar takes a little longer with payback times of 4-12 years. The warranties can be as long as 25 years for solar (5-10 years for wind) with the expected life being many more years beyond the warranty. Installing this type of equipment offers true **"Energy Independence."** No more electric bills, no more natural gas heating bills, and no cost to fuel your vehicles for all your local driving. This is not a pipe dream – it is possible, it is necessary, and it is only several thousand customers away from reality.

Making such a home hydrogen system affordable is one of my personal objectives, and I hope many other companies and organizations are pursuing the same goal, providing competition to further lower prices on such systems. See Chapter 10 to find out how you can do your part to make these systems more affordable.

DISTRIBUTED GENERATION
One of the more exciting mechanisms to reduce energy consumption is to build small renewable energy projects in communities where the energy is consumed. As opposed to the grossly inefficient current design of centralized power plants sending energy over hundreds of miles with associated losses, distributed generation eliminates nearly all the losses and inefficiencies by producing energy near where it is used. Examples are small wind plants that can power entire

---

[128] San Diego Gas and Electric.
[129] From ClimateStar.org, a partnership of the Union of Concerned Scientists and the Earth Communications Office.

towns, waste-to-energy facilities that convert residential trash, sewer waste, and other municipal waste into energy (usually in the form of either pathogen-free pellets that can be burned with low emissions in a generator or production of methane, hydrogen or ethanol), municipal solar plants, and landfill gas plants. A terrific option for the 50,000 coal miners that would be displaced by The Freedom Plan is to retrain and employ those hardworking coal miners in distributed generation plants in towns and cities nationwide. The author estimates that at least 300,000 new jobs would be created by The Freedom Plan at distributed generation plants alone.

The combination of energy efficiency in homes and commercial buildings along with distributed generation can significantly reduce energy consumption. This both reduces energy demand quickly and cost-effectively nationwide and nearly eliminates the need for new power plants. This allows the other steps of The Freedom Plan to truly convert all our energy generation to clean power.

STEP 4

Existing power plants are killing people – literally – with their pollution and emissions (see Chapter 3). Therefore, a major objective of The Freedom Plan is to rapidly begin converting these dirty, inefficient plants to cleaner, renewable sources of power.

COAL-FIRED POWER PLANTS

Coal plants are by far the worst pollution culprits (there is no such thing as clean coal technology – it's merely a phrase coined by the coal industry to try to get us to tolerate the construction of more coal-fired power plants). These deadly coal plants must be shut down or converted to renewable power. Biomass (wood waste, or energy crops such as buffalo grass, switch grass, mustard seed, rape seed, hemp, etc.) should be the only material burned in coal plants. Some existing coal plants today are already blending biomass with coal to reduce emissions, lower energy costs, and support the local rural community. Many other coal plants were built decades ago and meet one or more 'grandfather' clauses, allowing them to operate without modern emissions reduction equipment.

However, even the most modern coal plant in the United States, the Hawthorne plant in Kansas City, Missouri (owned and operated by Kansas City Power and Light) has failed its emissions requirements. When it was rebuilt in 1999 (after an explosion destroyed part of the plant), it won numerous awards for its emissions controls, getting national recognition for its modern 'clean coal' technology. As of June 2004, the plant has repeatedly failed its emissions tests according to an article in The Kansas City Star, spewing far more nitrogen oxide and sulfur dioxide than its emissions control systems were ever supposed to allow. Yet this time it is not making national news. "This is a major violation" of 2004 air quality standards says Lisa Hanlon of the EPA. The utility and the Environmental Protection Agency (EPA) will likely take up the fight in court over potential fines assessed by the EPA and/or the time allowed for the utility to fix the problem to meet the plant's required emissions targets.

It is fortunate that the EPA is watching the emissions from coal-fired power plants however, astonishing as it may be, the most toxic and deadly pollutant of all – mercury – is not even regulated by the EPA. Therefore, coal-fired power plants all over the nation can spew as much deadly mercury as they wish into the air. Mercury collection systems are only capable of extracting a maximum of 16 percent of the mercury according to EPA officials from the Kansas City office. The rest goes right up the stack and out the plume, traveling for hundreds of miles and settling into our lakes, rivers, streams, and ecosystems. As discussed in Chapter 3, just 1 gram of mercury in a 25-acre lake poisons all the fish in that lake making them unfit to eat by children or women of childbearing age.

It may shock you (and make you mad) to learn that as recently as July 2004, there were 106 new coal-fired power plants being planned for construction. Why? Three reasons:

1. Electric demand grew steadily in the 1990s (due to the Internet explosion that created a need for more computing power, steady population growth, increased standards of living, etc.).

2. Because people perceived (correctly) that coal-fired power plants are dirty and unsafe, this demand was met with a large growth in natural gas-fired power plants. Coal plants are also more costly to build than natural gas plants, they take longer to permit, and natural gas is somewhat cleaner than coal (but still very noxious).

3. Because of the rapid growth of gas-fired power plants, the increased demand for natural gas (NG) has led to record prices and a natural gas shortage that is considered by many to be a crisis. The outlook for NG shows that the price will remain above $5 per Mcf (thousand cubic feet) for at least the next five years (for comparison, this price was $2 in 1990). Therefore, because gas-

fired plants are no longer as cost-effective, coal is getting more attention now as the only other alternative that the industry understands or is aware of.

Most of these coal plants will never be constructed because of increasing public awareness and well-managed opposition campaigns. But some of them will be built (a few were permitted years ago when awareness was lower), despite our best efforts to stop them. Some experimentation is certain with coal gasification, a promising but extremely expensive technology that extracts flammable gases from coal and burns those gases instead of the coal itself. This process is cleaner and more efficient than burning coal, but still creates huge quantities of carbon dioxide. It is also very costly, unproven, and the plants are expected to take nearly a decade build. It remains to be seen whether this technology will ever become cost-effective relative to wind energy and other low-cost but clean electric generation methods. And, injecting carbon dioxide into the ground instead of releasing it into the atmosphere makes many climate scientists nervous as the impact of this technique is largely untested as well.

Fortunately, new coal plants must meet modern emissions standards. But unfortunately, those standards still allow huge clouds of particulates (linked to heart disease, lung disease, asthma, headaches), sulfur dioxide (acid rain), nitrogen oxide (smog, ozone, exacerbates headaches and asthma), carbon dioxide (global warming), carbon monoxide (deadly gas, also exacerbates migraine headaches), and more to billow right up out the stack and travel for hundreds or thousands of miles. These toxins are reduced in modern coal plants, but are still poisonous enough to kill nearly every living thing in a small town if the cloud was concentrated enough above that one town. And, the worst of all – mercury – is not even on the list of regulated toxins at all!

The good news is it is likely that after a few more are built in the next few years, these will be the last coal plants ever to be constructed in the United States. At some point, no more new coal plants will go online in the U.S.; probably the year 2010 (it takes years to permit and build a coal plant).

However, the deadly coal plants that are killing Americans and making the rest of us sick will still remain operational. WE MUST SHUT DOWN ALL EXISTING COAL PLANTS, starting with the oldest ones first, or convert them immediately to cleaner, renewable fuels. Adequate electricity generation must exist before we can begin shutting them down. Hence the need for massive amounts of new wind power to allow us to scale down coal-fired generation over a 5-year period. All coal plants in the U.S. could realistically be converted to biomass or shut down altogether by the 10th year of the plan (see Table 1 "Timeline").

We burned coal for heat and power in the 1800s. 200 years ago, we began using coal, digging it up out of the ground, hauling it around on train cars, and then shoveling it into boilers to create steam to turn an electric turbine. *We're still doing the same thing today, 200 years later!* The only difference is we have mechanized equipment doing the digging and shoveling, but the basic concept is still two centuries old. It is simply not necessary any more. Those who say we need it – those who are desperately clinging to this archaic and poisonous energy source – are doing so for a) money (they are in the coal industry and receive direct or indirect benefits from its use and existence) or b) because they are afraid to try something new. I have heard the following phrase from utility managers and executives countless times: "Well, this is how we've always done things around here."

ATTENTION COAL INDUSTRY EXECUTIVES:

Don't worry about your job: we can put you and your staff to work converting America to clean, renewable fuels. And someday, a truly clean, pollution-free form of power generation using coal as its feedstock may be available.

GAS-FIRED POWER PLANTS

Converting natural gas plants to burn renewably produced methane, Hythane, and hydrogen can be achieved with even greater ease than converting coal-fired plants. As the injection of gas sources occurs during Step Three, gas-fired power plants can be modified as needed to function using these cleaner, renewable fuels. Also, each gas plant should be required to add the most modern emission control systems available, and should only be used in emergencies for backup power or for peak shaving during periods of very high electric demand (July and August when large numbers of air conditioners are running).

To convert a modern gas-fired power plant to burn renewably produced methane takes no modification at all. To allow it to burn Hythane, some minor modifications may be necessary if the percentage of H2 exceeds 15-20 percent of the blend. Only when the gas turbine is burning pure hydrogen or a hydrogen-rich blend of Hythane are major modifications required. This would include new fuel injection controls, stainless steel exhaust (due to the high water content of the exhaust) and a converter or turbo-charger device to minimize NOx emissions. The conversion timeline for these activities will facilitate itself as Step Four unfolds.

STEP 5

The final step of the Freedom Plan is the most exciting one. Now that we have powered the entire electric grid with clean wind and solar power, renewably produced hydrogen and methane, and biomass, we can focus on renewable fuels for our cars, trucks, buses, trains and airplanes.

CLEANER FUELS

Some vehicles already use E85 (85 percent ethanol, made from corn) or bio-diesel (made from soybeans or recycled cooking oils) is discussed in Chapter 2. And some use compressed natural gas (CNG: Ford Motor Company introduced many CNG vehicles in the late 1990s, and Honda Motor Company currently sells the CNG-powered Civic GX). These fuels are cleaner and often safer than fossil fuels, and are becoming more and more popular. For example, the Evo Limo Service in Hollywood, California has converted its entire fleet of limos to run on CNG, lowering smog-forming emissions by nearly 80 percent over gasoline.[130] This limousine service is so popular with some movie-stars that it is the only company they call any more.

Demand for hybrid vehicles have also grown quickly. The Toyota Prius is so popular that by the summer of 2004, dealers were no longer accepting names for the waiting list. The Ford Hybrid Escape SUV (the world's first hybrid SUV, with gas mileage of over 35 MPG in the city) had 35,000 units sold (out of a total first-year production of only 20,000) before the 1st one was officially sold at a dealership. The Honda Insight and Civic hybrids have also exceeded all sales expectations. Many more hybrids are on the way.

Because Steps 1-4 will build out a clean electric grid and a hydrogen pipeline infrastructure, the installation of CNG, Hythane and H2 fueling stations at gas stations nationwide will occur naturally. The costs are reasonable since existing NG pipelines are nearly ubiquitous, connecting homes, neighborhoods, business parks, strip malls, shopping centers, and nearly everything else in urban and suburban areas. General Motors estimates it would cost $12 billion to modify gas stations to also sell hydrogen fuel at 12,000 sites in 100 of the nation's biggest cities[131] (a cost estimate that is in line with figures contained in The Freedom

---

[130] Stefan Lovgren in Los Angeles for National Geographic News, April 19, 2004
[131] USA Today, Money section cover story "Alternative-fuel vehicles star, but wide use is miles away" January 12, 2005.

Plan). Rural areas will have access to renewably produced propane, methane and hydrogen delivery services.

Ethanol and bio-diesel fueling stations are even easier to install, since the transport of those fuels is significantly easier than hydrogen. Therefore, as these fueling dispensers offering H2, Hythane, CNG, ethanol and biodiesel become more commonplace, automobile manufacturers (as well as start-up entrepreneurial companies making kit-cars, motor scooters, motorized bicycles, and miniature errand-runner cars) will begin producing the vehicles that operate on those fuels. As more are produced, costs will decline.

ENVIRONMENTALLY FRIENDLY CARS

Progressive automobile companies will begin making lighter cars with carbon fiber and composites (like those used in aviation) that require less power (smaller, more efficient motors) due to their lower weight. In fact, plans exist for a light carbon-fiber vehicle that is stronger and safer than steel, only about 20 percent of the weight, and made using carbon fibers from landfills: a totally renewable and recyclable car[132]. Carbon fiber body panels are impervious to door dings, have the color blended into the fiber (no painting needed, nor would paint chipping be a problem), and could be easily interchanged if you want a new color. A car of this design would get well over 100 miles to the gasoline-gallon equivalent if it was powered with a hydrogen fuel cell (running at 60 percent efficiency, compared to the best modern internal combustion engine which runs at only 25 percent efficiency) and a small electric motor with regenerative brakes with a thin-film solar PV array on the roof. Imagine filling this car up with free hydrogen you make at your own home from rainwater and solar energy, and being able to drive a thousand miles without refueling with 10 kilograms of H2 stored in safe carbon fiber wrapped tanks.

In addition to a plethora of newly designed clean fuel vehicles that would be readily introduced, existing vehicles could be converted to run on these cleaner fuels, especially hydrogen. A complete conversion kit for a standard engine on any car would be $2,000 or less when mass produced, and will only take a couple of hours to install (the same amount of time as an engine tune-up). The most expensive component of the hydrogen conversion is the high pressure, impact resistant storage tank. 5,000 psi tanks today cost about $5,000 each and 10,000 psi tanks cost over $40,000 each. This is because they are custom-built mostly by hand. As mentioned earlier, the cost of the tanks and other components will

---

[132] Courtesy of the Rocky Mountain Institute: http://www.rmi.org/sitepages/pid386.php.

drop considerably with mass production (even with an order for as few as 100,000 vehicle conversion kits, the costs will drop enough for many early adopters to affordably convert their cars to hydrogen).

YOUR HOME IS A CLEAN POWER PLANT

With enough participation, the costs of a turnkey residential power plant (as described in Step Three) will drop to $25,000 or less. This will produce all your electricity, hot water, heated air, and fuel for up to two of your vehicles' local driving needs, <u>with almost no additional expense</u>, for the next **50 years**. It includes a rooftop solar power array, rainwater collection, hydrogen-maker (about the size of a footstool), fuel cell (about the size of a small end table), and a hydrogen fueling dispenser (the size of toaster with a hose coming out of it). These should be readily available from some suppliers by the year 2006, with costs dropping to $10,000-$15,000 by 2015.

Many will choose to lease or finance these systems, in some cases using a home equity loan to be able to deduct the interest payments on their tax return. A typical cost might be $100-200 per month to finance a $15,000 system over 20 years at a low interest rate (remember the lifecycle of all but the fuel cell is 30-50 years, and the fuel cell membrane can be replaced like a printer toner cartridge for $1,000-$2,000). Hopefully, low interest loan programs from large financial institutions and government-backed loans will materialize to support widespread installations of these systems.

Commercial systems (for businesses, office buildings, hotels, etc.) will be even more cost-effective than residential systems due to their size, so these will no doubt be produced by a variety of manufacturers. In fact, some of the companies that will excel at producing fuel cells, solar roofing materials, and hydrogen-makers (electrolyzers and other methods of producing H2) have not even been created yet. Some of people who will think up the industry-leading ideas have not even heard about clean power yet. As the word continues to spread about the damage we're doing to ourselves and our home (Earth), the entrepreneurial spirit that made this country famous will energize these people (pun definitely intended) to drive the clean power crusade forward.

INVENTORS WANTED!

Over the next 10 years, I believe we will witness one of the most exciting times this world has ever seen. The growth in technology over the prior 10 years has been mind-blowing, doubling every six months. With all the new technology and its critical computing power comes increased electrical demand, which has exacerbated the problem of relying on fossil fuels for our energy.

During the next 10 years, individuals of all walks of life will have an amazing opportunity to contribute to a time period that will be even more mind-blowing: converting the largest industry in the world by far (the energy industry) to clean, renewable, safe and healthy power.

To accomplish this enormous feat, we need the brightest minds and the most creative thinkers to accelerate this Freedom Plan, and lower the costs wherever possible to economically force fossil fuels out. The more financial sense it makes to convert (even without including the costs of health and environmental damage), the faster we can get the 'old school thinkers' to convert.

There will likely be dozens of new inventions and developments that can be implemented by the architects of The Freedom Plan. Areas of interest that already show promising results are bacteria that can digest spent radioactive waste from nuclear power plants and generate electricity at the same time,[133] bacteria that create hydrogen while eating landfill or municipal waste, carbon nanotubes for storage of hydrogen in compact containers, and others.

The sheer size of the enormous energy industry boggles the mind. If you add up the worldwide revenues of the entire computer industry, then add up the worldwide revenues of the entire telecommunications industry, you'll have a sum that is approximately 10 percent of the energy industry. *In other words, the energy industry is 10 times bigger than the computer and telecommunications industries combined.* Therefore, inventors and entrepreneurs who usher in the Renewable Age stand to gain enormous profits: the energy pie is huge, so even a tiny slice can be worth billions.

ECONOMIC IMPACT

A 2001 report shows that California would gain 141,400 jobs by 2020 by implementing more clean energy policies, especially energy efficiency policies.[134] Apply that impact to the rest of the country and we find that over three million new jobs would be created. However, in the Midwest where huge quantities of new wind, transmission line, biomass, hydrogen projects will be built should conservatively add another two or three million jobs. Keep in mind, there are

---

[133] SCIENCE Magazine, December 12, 2003.
[134] Tellus Institute and MREG & Associates

hundreds of thousands of people across the globe currently working for companies supplying the U.S. with our oil and fossil fuels. Nearly all of those jobs will come back home to America. This alone has a tremendous economic impact to our nation's wealth and prosperity, not to mention energy security.

Wind technologies also increase the American job outlook: they offer more jobs per unit of energy produced than other forms of energy. With the right investment, wind energy facilities could create 1.7 million jobs worldwide by the year 2020 and almost a million jobs in the U.S. alone over the next 12 years.[135] That assumes only a modest 5-10 percent of our nation powered by the wind, let alone the 100 percent called for by The Freedom Plan.

The "New Apollo Project" coalition study indicates 3.3 million new hi-wage construction and manufacturing jobs would be created by 2015 with a commitment to more renewable energy in America. I believe this estimate is accurate with about 20 percent of our energy coming from renewable sources. At 100 percent renewable, America will likely see 8-10 million new jobs, which could lead to the greatest period of economic growth and prosperity our nation has ever seen.

A recent study by the Union of Concerned Scientists indicates that if the nation were to produce 20 percent of electricity from renewables, 50,000 jobs would be added in Texas alone, as well as $10 Billion in new capital investment and two-thirds of a billion in tax revenues for small towns and school districts would be added in Texas.[136] The economic impact for the entire nation would easily be 20 times that amount for a 20 percent renewable goal, and 100 times that amount for a 100 percent target. That total is 5,000,000 new jobs, $1 Trillion in new capital investment, and $166.25 Billion in tax revenues (over 50 years). And that is only for electricity generation. Add the new jobs for producing our all transportation fuels from renewables and you can see how quickly the numbers can add up.

According to the Iowa State Input Output model for wind energy, as discussed earlier in Chapter 4, every megawatt of wind energy installed creates $3 Million of economic impact over 20 years. With over two million megawatts included in The Freedom Plan, that represents an economic impact of a whopping $6 Trillion over 20 years. That figure, which is three times the size of the entire annual budget of the U.S. government, is just the positive economic impact from wind. Add the hydrogen components, transmission lines, biomass, renewable methane

---

[135] GRACE (Global Resource Action Center for the Environment) Green Energy Project, New York, NY at http://www.gracelinks.org/energy/wind/.

[136] Business Wire, 10/5/04, "TX Leads Nation in Air Pollution from Coal Plants," Austin.

and solar installations to the equation and America could enjoy well over $20 Trillion of positive economic benefits over 20 years, or $50 Trillion over the life of most of the clean power equipment installed as part of The Freedom Plan. What does that do to the argument that reducing greenhouse gases will hurt our nation's economy?

CONCLUSION

America was founded on principles of freedom. One freedom most of us do not have is freedom from the power grid and utility companies. It will soon be within reach, if and only if enough people take action to make it possible. The government will NOT bring you energy independence – ever. If the reader is hoping to eventually elect the right government to rescue us from the shackles of the energy industry, you will die waiting for it. Regardless what political affiliation the reader has, it is not possible for a reactive (not proactive) government to save us from ourselves. Too much money and too many interests are at stake for ANY President, let alone an entire political party, to 'throw coal to the wind' and fully embrace such a massive and necessary conversion. The only way the government will take action faster is if a major and sudden disruption in our oil supply were to occur. Even then, the inherent interests in Washington D.C. will ensure that a huge emphasis is placed on drilling for more fossil fuels domestically, and find new uses for coal, coal-bed methane, and other dirty energy sources. Even if a President is elected that thumbs his or her nose at powerful energy industry political donors and embraces the Freedom Plan (or another similar solution), you can forget about getting a supermajority of Congress to agree on this issue. The only way we can convert our nation to clean power is using grass roots marketing and education to follow a well-designed, cost-effective plan using market mechanisms – not relying on any government administration to make it happen.

Year	MW of Wind Power	Power Lines	MW of Electrolyzers	% of NG produced renewably	% of H2 in NG system	% of Coal Plants Converted to Biomass or Closed	% of Homes producing 100% of their Energy needs
1	2,325	5					
2	32,985	10	100	1%		1%	
3	40,816	15	500	3%		3%	1%
4	40,740	20	5,000	5%	1%	5%	2%
5	56,825	30	25,000	10%	5%	10%	3%
6	87,314	30	100,000	20%	10%	30%	5%
7	184,727	30	300,000	30%	20%	50%	10%
8	428,523	10	500,000	50%	40%	70%	20%
9	728,449	10	750,000	75%	60%	90%	30%
10	1,138,705	10	0	100%	80%	100%	40%
Total	2,741,410	170	1,680,600				

Therefore it is up to us, as freedom-loving American citizens, to force the conversion of our great nation to cheap, renewable, clean power sources. We owe to our health, our families, our children, their children and ourselves to do our part. It is the right thing to do, it feels great supporting such a true and noble cause, and as you'll see in the next chapter it will save us a bundle.

Chapter 7:

SAVE $20 TRILLION OVER 20 YEARS

The Freedom Plan described in the previous chapter is possible, and it is necessary. But it will be extraordinarily difficult to make a reality. Too many vested interests will resist such a revolutionary change in the way we procure our energy. Although I may be considered an environmentalist by many readers, I consider myself a realist. And I am not so naïve that I think we can accomplish this conversion without broad support from a vast majority of the public. To that end, I have done a tremendous amount of research on the cost of our current energy policy to American taxpayers. As a result, I wish to make a compelling case for supporting the implementation of the Freedom Plan – from a purely economic standpoint.

Why? This is because we need both the 'Green' environmental camp and the politically conservative camp to embrace the Freedom Plan. I am a political conservative who believes a small, efficient government is preferable to a large, bureaucratic, invasive government that regulates too many aspects of our lives and/or business community. Therefore, I feel it is critical to be certain this conversion makes sense financially. We must be able to make a compelling case that The Freedom Plan will *enhance* our economy, not damage it. And it must not simply increase the size, expenditures and regulatory scope of our already bloated and inefficient federal government. The Freedom Plan <u>must</u> be driven by consumer demand to be effective. And that is exactly why it will succeed. But the message that such a plan exists must reach the masses.

So, to build the financial case of why clean power is so critical this chapter lists aspects of current U.S. energy policy and demonstrates the cost to American taxpayers over 20 years – if we continue down our current path and do not implement the Freedom Plan. This allows a fair and reasonable comparison to be made between the status quo (continuing to rely on fossil fuels as the mainstay of our energy policy) and implementation of the Freedom Plan. A case will be made using sensible projections and real data, always using conservative estimates wherever possible to weaken any potential arguments from critics.

At the conclusion of this chapter, it is my hope that the case I have tried so hard for three years to make for The Freedom Plan not only appears plausible to you, but also feels right. And finally, I hope to create a sense of urgency for you to do your part to spread the word about this attractive solution to one of our most pressing problems. America needs to change her energy policy; perhaps after reading this book you may even help to spread the word about this Clean Power Revolution that is slowly gaining momentum.

THE END OF CHEAP OIL (Savings – $5.5 Trillion)

According to Time magazine[137] United States oil imports account for over 62 percent of our total consumption today, and are expected to rise to over 80 percent by 2025. They actually understated our dependence somewhat since they were relying on 2001 data and did not include some imports of other petroleum products; we actually import closer to 72 percent of our oil. In 2003 that meant we sent **$164 Billion** to foreign countries to satisfy our thirst for oil and petroleum products. This cost is based on oil at $29 per barrel, total oil consumption of 21 million barrels per day, and oil imports of 15.5 million barrels per day (72 percent).[138] Since oil prices hit all-time record highs during 2004 (surpassing $55 per barrel in 2004 and hovering around $48 at the time of this writing) that figure using $29 oil is likely to be grossly understated going forward. We will undoubtedly send even more of our American greenbacks overseas to ship more oil to our thirsty vehicles and oil-fired generators.

This dependence on foreign energy costs our economy more than many people realize: every dollar we send outside our borders to buy products or services such as oil lowers our national wealth by a dollar. This trade imbalance can, over time, severely hamper our nation's ability to finance our economy. We have been importing tens of billions of dollars of oil every year for decades, and it is starting to show. Our national savings rate is too low, our capital is flowing to countries and to people that do not like us, our trade deficit is at an all-time high, our federal budget deficit is at an all-time high, and there is not as much cash leftover in the U.S. financial system as there should be to reinvest in American businesses.

---

[137] TIME magazine, July 21 2003, "Why the U.S. is Running Out of Gas."
[138] Scripps Howard News Service "Global Oil Production & Consumption" by Joan Lowy, October 28, 2004 [sources: EIA, International Energy Agency, U.S. Geological Survey]

Roughly 2/3 of known oil reserves are located in the Persian Gulf region, primarily in Saudi Arabia. It may come as a surprise to learn that in the early 1900s, the United States has oil reserves so large that we actually had as much oil as Saudi Arabia. And, the U.S. was the world's largest oil producer for most of the last century until the 1970s when our reserves began to dwindle. As of 2004, the U.S. has only 2.5 percent of the known reserves and yet we are the largest consumer of oil in the world by far, using about 25 percent of all the oil consumed in the world[139] with only 4 percent of the world's population.

There are three primary issues that affect our continued dependence on oil, which will likely make this dependence progressively worse:

The price of oil will continue to remain high and trend higher.
U.S. demand for oil continues to rise.
The oil we import from overseas will continue to increase as a percentage of the total oil we consume (from 72 percent today to as high as 85 percent by 2025 by some estimates).

Therefore, to attach a cost to our continued dependence on oil, we shall address each of these three issues to calculate the overall cost to all Americans of our continued dependence on oil.

Stephen Leeb, president of Leeb Capital Management and co-author of "The Oil Factor" says crude oil prices will likely strike **$100 per barrel** by the end of the decade if not sooner.[140]

Since $29 oil is a thing of the past, we must establish a more reasonable price per barrel going forward of $40-100. Most analysts do not expect to see sub-$30 oil for an extended period of time – ever again. And they suggest that we will surpass $100 oil in the near future. There are many factors to support this assertion, not the least of which is rising worldwide demand and tightening supplies. China leapfrogged Japan as the world's 2nd largest consumer of oil in 2003, and is expected to increase consumption from two million barrels per day (a mere 10 percent of U.S. consumption) to over ten million barrels per day by 2025 (50 percent of current U.S. consumption).[141] Other developing nations in Asia and elsewhere are rapidly industrializing, requiring still more oil to feed their growing economies. Barring an enormous increase in oil prices (above a

[139] National Geographic magazine, "The End of Cheap Oil" June 2004 p.85.
[140] "Not Many Alternatives for Energy Investors" by Meg Richards, AP, NY, 10/2/2004.
[141] National Geographic magazine, "The End of Cheap Oil" June 2004 p.89

threshold of about $200 per barrel), worldwide oil demand is expected to continue rising steadily.

Many people do not realize how greatly oil permeates our lives. For example, to raise a steer to full weight of 1,250 pounds, it takes six barrels of oil (283 gallons). That includes fertilizers on corn fields to diesel that runs the farm machinery and everything else needed to raise that beef steer. In other words, every pound of beef takes ¾ of a gallon of oil to produce.[142] The tires on your vehicle each contain seven gallons of oil, used in their manufacture. Oil is used to make medical implants, fertilizers, computers, bike helmets, and nearly all things plastic: children's toys, tennis shoes, polyester clothing and filler products for pillows and comforters, automobile parts, appliances pieces, and nearly everything else you can think of with plastics or polymers. Once we locate and develop an alternative to oil to drive the energy industry, we will be able to use oil for more valuable products such as those mentioned.

In the U.S., there is controversy over drilling in Alaska's Arctic National Wildlife Refuge. It contains an estimated 10.4 billion barrels of technically recoverable oil,[143] which is the largest remaining reserve of oil by far in the U.S. However, if we decided to drill there and then recovered every drop, it would not make much of a difference. At today's consumption levels of 22 million barrels per day in the U.S., it would only last 15 months.

Other disturbing trends:

Most estimates of total recoverable oil worldwide since production began range from about 1.8 trillion to 2.4 trillion barrels. The U.S. Energy Information Administration's estimate is overly optimistic at 3 trillion barrels. About 950 billion barrels have already been produced. Therefore, at current worldwide consumption of 82 million barrels per day and using 2.4 T bbl as the total, **we will completely run out of oil in 48 years.** This assumes we do not increase daily consumption (we will) and that we can recover all of the 2.4 trillion barrels estimated (the cost continues to increase as the more hard-to-reach sources have to be tapped).

The funds that oil companies allocate to research have decreased more than 50 percent since 1990. The membership of the American Association of Petroleum Geologists peaked in May 1986 at 44,757 members and is now less than 30,000 members.

---

[142] National Geographic magazine, "The End of Cheap Oil" June 2004 pp.98-99
[143] US Dept of the Interior report 3/12/03 "ANWR Oil Reserves Greater Than Any State"

"I think in total the outlook is much too high for production and it is unrealistic for the world to be expecting such high numbers from all of the producers, including Saudi Arabia. They're not only overestimating the Middle East, but they overestimate non-OPEC, they overestimate Russia, they overestimate the whole global resource base. And I think this is a rather dangerous situation for the US government policy to be based on." Sadad al-Husseini, former vice-president of **Saudi Arabia's national oil company Aramco**, covered on Channel 4 News, 26 October 2004.[144]

"Your wealth is endangered": a deep world economy crisis of unprecedented size is inevitable, unless effective countermeasures are immediately carried through. This is the consequence of multiple global problems, the most important ones being: global warming predicted to cause unexpectedly rapid climate change, unexpectedly rapid depletion of global oil resources, ongoing and accelerating mass extinction of species and increasing overpopulation. This will have serious consequences unless the course of events is drastically changed. Because of the nature of this process, measures have to be taken immediately. When the severity of the situation becomes even more obvious, it will be much too late to do anything about it. Even now it is becoming late....[145]

New discoveries of oil fields per year reached a total of zero for the first time in history in the year 2003. "We now consume six barrels of oil for every new barrel we discover. Major oil finds (of over 500m barrels) peaked in 1964. In 2000, there were 13 such discoveries, in 2001 six, in 2002 two and in 2003 none."[146]

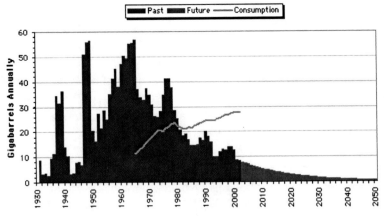

---

[144] http://www.btinternet.com/~nlpwessex/Documents/energycrisis.htm
[145] Urgent message to the wealthy From Physicians and Scientists for Responsible Application of Science and Technology (PSRAST) Published at February 20, 2004
[146] "Break out the bicycles" Guardian, 8 June 2004

Oil Discovery (3 year average - past and projected) 1930-2050
Source: Association for the Study of Peak Oil

"Civilization is at a turning point. **In the next 50 years, we will experience the biggest surge in energy demand in history.** Yet a growing number of experts are warning that the rate at which we are able to pump oil from the ground is likely to peak within a few years. If we want to keep our cars on the roads and the lights on in our homes, we need to find a new source of energy and fast..."[147]

"Our industry can certainly be proud of its past achievements. Yet the challenges we will face in the coming years will be every bit as great as those encountered in the past, due in part to ever-increasing global energy use. For example, **we estimate that world oil and gas production from existing fields is declining at an average rate of about 4 to 6 percent a year.** To meet projected demand in 2015, the industry will have to add about 100 million oil-equivalent barrels a day of new production. That's equal to about 80 percent of today's production level. In other words, by 2015, we will need to find, develop and produce a volume of new oil and gas that is equal to eight out of every 10 barrels being produced today."   John Thompson, President of **ExxonMobil**, the world's largest oil company.[148]

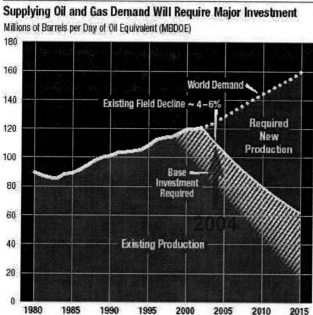

<hr />

[147] Power Struggle, NewScientist, 2 August 2003 (print edition p8)
[148] The Lamp (published for ExxonMobil shareholders), 2003, Vol. 85 No.1

"The Clean Power Revolution"
Graph from ExxonMobil report 4 February 2004, p4 (2004 marker added for illustration)
'A Report on Energy Trends, Greenhouse Gas Emissions, and Alternative Energy'

Wall Street Journal: demand for oil is expected to increase by about 2 percent per year according to the International Energy Agency, putting total demand at about 100 million barrels of oil per day by 2010 and 125 million barrels per day by 2025. "The world will not be able to produce more than 100 million barrels of oil a day. A peak in petroleum output is looming in the years ahead…just after 2010, and then begin a long-term decline [of oil production], the PFC Energy study found."[149]

FORTUNE Magazine: "With oil capacity stretched thin and supply fraught with uncertainty, the U.S. is more vulnerable than it was 30 years ago, when the oil crisis first traumatized the nation. Fortune's plan consists of four approaches: 1) Improve fuel economy (part of the Freedom Plan), 2) More spending on alternative fuels (part of the Freedom Plan), 3) Redoubled commitment to efficiency (part of the Freedom Plan), and 4) Getting serious about solar and wind (of course, part of The Freedom Plan). Some innovative companies are already taking leadership roles (like Krystal Planet, see Chapter 9) and it would be ideal to rely solely on market mechanisms (exactly what is proposed by The Freedom Plan). **If we do nothing, you can be sure Americans will pay more than just the price at the pump.**"[150]

"For the world as a whole, oil companies are expected to keep finding and developing enough oil to offset our seventy one million plus barrel a day of oil depletion, but also to meet new demand. By some estimates there will be an average of two per cent annual growth in global oil demand over the years ahead along with conservatively a **three per cent natural decline in production from existing reserves. That means by 2010 we will need on the order of an additional fifty million barrels a day. So where is the oil going to come from?** Governments and the national oil companies are obviously in control of about ninety per cent of the assets. Oil remains fundamentally a government business. While many regions of the world offer great oil opportunities, the Middle East with two thirds of the world's oil and the lowest cost, is still where the prize ultimately lies, even though companies are anxious for greater access there, progress continues to be slow." **Dick Cheney**, while Chief Executive of Halliburton, (later elected to be the Vice President of the United States).[151]

---

[149] September 9, 2004 from "Demand for Oil Could One Day Outstrip Supply."
[150] Fortune Magazine, August 11, 2004 "Plan to Free America from Dependency on Oil."
[151] Speech at London Institute of Petroleum, Autumn Lunch 1999

In summary, "People should be doing something now to reduce oil dependence," says Princeton physicist Alfred Cavallo, "and not waiting for Mother Nature to slap them in the face."[152]

The world today produces 82 million barrels daily; the petroleum experts say we will not be able to increase that production beyond 100 million barrels per day. Therefore, as demand increases, supply will shrink ever faster, exacerbating the problem.

In the interest of being conservative for the purposes of this economic cost estimate, I suggest that we use $40 per barrel, the lowest end of the price range provided earlier of $40-60 per barrel over 20 years. This provides an accurate, but probably understated, cost estimate to our economy of foreign oil dependence. So, assuming price of oil never rises above $40 per barrel for the next 20 years (an unlikely scenario); America would import $226 Billion per year worth of oil.[153] That represents a cost of at least $4.5 Trillion by 2025.

Next, we need to consider **increasing overall demand** for oil. The figure above assumes static demand and a fixed percentage of oil coming from overseas, both of which are certain to rise. According to the U.S. Energy Information Agency (a division of the Department of Energy), oil consumption by 2025 will be 28.3 million barrels per day.[154] Using that figure as the peak in 2025, and 21 million barrels today, average daily consumption should be at least 24.65 million barrels per day. This increased demand bumps our oil dependence cost up to $259 billion per year, or at least $5.2 Trillion by 2025.[155]

Finally, as U.S. oil reserves continue to dwindle, most analysts project that our reliance on foreign sources for oil and petroleum products will increase from 72 percent today to a range of 80-85 percent by 2025. Assuming the lower figure of 80 percent is realized, that provides an average over 20 years of 76 percent, bumping the $5.2 Trillion up to **$5.5 Trillion** by 2025.

Obviously, most people and energy analysts assume that oil prices will probably average more than $40 per barrel, demand may increase more than 7.4 percent over 20 years (that's only 0.4 percent per year), and consumption from foreign sources may soar to well over 76 percent. For example, just a $20 increase in oil prices per barrel bumps the cost from $5.5 Trillion to a whopping **8.2 Trillion** by

---

[152] National Geographic magazine, "The End of Cheap Oil" June 2004 p.96
[153] 15.5 million barrels per day x $40 x 365 days per year = $226 Billion per year.
[154] http://www.eia.doe.gov/oiaf/aeo/gas.html
[155] 24.65 million bpd x $40 x 365 days x 72% imports = $259 Billion per year.

2025.[156] A \$20 increase in average oil prices occurred in a matter of months in 2004.

No matter how you slice it, our dependence on oil costs our economy a fortune. If we could keep those dollars here in the USA, imagine how powerful a stimulant that would be for our economy. If just a portion of those dollars went to pay for The Freedom Plan, it could pay the entire cost to convert our nation to clean power. And, the cost of our energy would become the lowest in the world by far since it would all be driven by inexhaustible supplies of clean, renewable, wind, solar, and biomass power.

Minimum Cost to America through 2025: **\$5.5 Trillion**.

(More realistic cost to America through 2025: \$8-15 Trillion)

NATURAL GAS (Savings – \$4.2 Trillion)
According to the National Petroleum Council (a leading advisory service to the U.S. government on energy policy for over 60 years), the United States natural gas industry is facing two major challenges. Their report, issued September 25, 2003, states that American consumers will pay an additional cost of \$2.2 Trillion[157] over the next 20 years – over and above what we're already expecting to pay for natural gas. This is strictly an increased cost to the U.S. economy due to our current reliance on natural gas for heating and electricity. Their report cited two primary factors causing this increased burden on our society:

1. Continued volatility of natural gas prices, making energy planning more difficult and costly (cost: **\$1 Trillion** by 2025).
2. The need to obtain new supplies of natural gas, which will include imports of LNG (liquefied natural gas), imports from Canada, and increased domestic natural gas exploration and drilling (cost: **\$1.2 Trillion** by 2025).
3. Total additional cost to America (savings with The Freedom Plan): **\$2.2 Trillion** by 2025.

You may remember the all-time record highs for natural gas prices during the price spikes in 2001 and 2003 – if you use natural gas to heat your home you

---

[156] 24.65 million bpd x \$60 x 365 x 76% = \$410 B/yr x 20 yrs = 8.2 Trillion.
[157] National Petroleum Council report on http://www.npc.org/reports/ng.html.

probably remember those winter bills all too well. Most experts agree that natural gas prices will remain at current levels ($4 to $5 per thousand cubic feet) or higher for at least the next five years. In September 2004 before the winter heating season even began, natural gas prices jumped 17 percent in just two days to nearly $7 per thousand cubic feet, and the winter of 2004-2005 is expected to bring gas prices of $10-12 per thousand cubic feet, which would be a new all-time record high.[158] "It's inconceivable to me that...we'll ever see a return to $2 natural gas," said David Costello, economist with the federal Energy Information Agency.[159]

Projections for domestic natural gas consumption in 2025 range from 29.1 to 34.2 trillion cubic feet, compared with 22.6 trillion cubic feet in 2002.[160] Taking the average of the range of 31.65 trillion cubic feet represents an increase in demand of 71 percent, or 3.5 percent per year. North American natural gas demand by category is shown below:

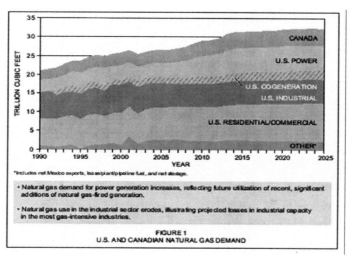

FIGURE 1
U.S. AND CANADIAN NATURAL GAS DEMAND

Source: National Petroleum Council

Increasing demand is not the only reason natural gas prices are increasing. Dwindling supplies are another major factor. The largest natural gas field in the world is the Hugoton field in western Kansas – its output has been steadily declining and is expected to run dry within 7-10 years. Other gas fields across the U.S. are seeing similar a drop off in production. Why? Because in the 1990s, the electric utility industry began switching their focus for new power plants

---

[158] USA Today "Natural Gas Supply Fears Raise Prices" by Barbara Hagenbaugh, September 30, 2004.
[159] Kansas City Star newspaper "The Heat is On" by Steve Everly, March 30, 2004.
[160] U.S. Energy Information Agency, http://www.eia.doe.gov/oiaf/aeo/gas.html.

away from dirty, controversial coal plants to cleaner burning natural gas-fired power plants. That has helped reduce some of the toxic emissions from electricity generation but has increased pressure on the already strained supply and distribution channels of natural gas. Here is where our natural gas comes from by source:

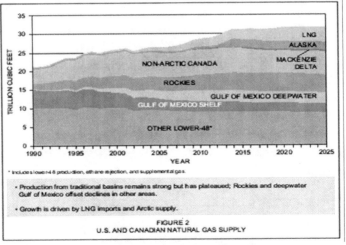

FIGURE 2
U.S. AND CANADIAN NATURAL GAS SUPPLY

Source: National Petroleum Council

One solution to the natural gas supply shortage is unfortunately a new dependence on foreign energy: liquefied natural gas (LNG) terminals were built in the 1980s after a jump in natural gas prices. Imports of LNG were modest until 2001 when the U.S. began importing LNG in large quantities for the first time in history. Federal Reserve Chairman Alan Greenspan said in 2003 that "LNG imports were needed to wring some volatility out of gas markets."[161] Recent attempts to build new LNG terminals (over 40 new such terminals are proposed across North America) at seaside towns across the U.S. have met with powerful resistance. Recent attempts in Maine, Massachusetts,[162] California, and Alabama to approve construction of LNG terminals were all soundly defeated due to a well-orchestrated public effort.[163] The public outcry against LNG terminals stems from fears that an accident (such as the one in Algeria in early 2004 that killed 28 people or the one in Cleveland Ohio in 1944 that killed 128 persons and injured 225 more) or terrorist attack could cause an explosive fire killing nearby residents. And, the public simply does not want to see America become dependent on yet another foreign source of energy. But North American

[161] Kansas City Star newspaper "The Heat is On" by Steve Everly, March 30, 2004.

[162] The Wall Street Journal "Opposition Greets Gas Terminal Plants" by John Fialka and Russell Gold, May 15, 2004.

[163] "LNG to the Rescue" by Steve Everly, Kansas City Star, October 5, 2004.

supplies, although still substantial, will simply not be able to meet demand over the next 20 years.

Over 70 percent of the world's natural gas reserves are located in the Middle East (33 percent of the total) and Russia and her satellite countries (40 percent of the total). The cost of building new plants to heat up the LNG from negative 260 degrees to its gaseous temperature can cost $500 million and up.

According to The Wall Street Journal, demand is rising in the U.S., but most North American supplies are flat or in decline, leading to rising prices and the growing risk of heating fuel shortages and blackouts.[164]

Since 1979 the price of natural gas has risen by 226 percent, compared with overall inflation at 159 percent according to the Bureau of Labor Statistics. One result of rising natural gas prices: 15 fertilizer plants have closed permanently in the United States, and imports are taking more of a share of the fertilizer market. Another result: bakeries across the nation, who rely on natural gas for their ovens, are being forced to increase prices of their baked goods to recover some of the increased energy costs. Average winter heating bills for the Midwest region (which are some of the lowest in the nation) are expected to be $993 during the 2004-2005 winter season, compared to $602 during the 2001-2002 season[165] (a 60 percent increase in just three years). Worries about additional price spikes over the next two decades are justified: "It's not going to take much to set off prices," said Mark Stultz, a spokesman for the Natural Gas Supply Association.[166]

By 2025, natural gas prices are projected to be about $8.50 per thousand cubic feet (Mcf). From 1999 through 2003, natural gas prices averaged $3.56 per Mcf. The average in our baseline year of 2003 was $4.98 and 2004 will be even higher. Therefore, the average of the 2003 cost of $4.98 and the projected price in 2025 of $8.50 is $6.74.[167] This price now appears to be a low estimate due to recent upward pressure on the price on several fronts, including resistance to new LNG terminals, low investment in infrastructure, and higher demand for natural gas for electricity generation due to record high oil and diesel costs and continued resistance to new coal-fired power plants to meet rising electric demand. An average natural gas price of $6.74 per Mcf over 20 years is $3.18 higher than what

---

[64] The Wall Street Journal "Opposition Greets Gas Terminal Plants" by John Fialka and Russell Gold, May 15, 2004.

[65] U.S. Energy Information Agency.

[66] Star Business Weekly "Winter Predictions include High Gas Bills" by Steve Everly, September 13, 2004.

[67] U.S. Energy Information Agency.

we are used to, on average, since 1999. With consumption of 31.65 trillion cubic feet per year until 2025, this results in an increased cost to the U.S. economy of $100 Billion per year or **$2 Trillion** over 20 years in 2002 dollars.

A study commissioned by the Colorado Public Utility Commission in 2001 found that when natural gas reaches or exceeds $3.50 per Mcf, producing electricity using wind energy is less expensive. This is true even when comparing wind-generated power to power from an existing gas-fired power plant, solely due to the fuel cost. In other words, the financial burden on Colorado consumers is less to let the gas-fired power plant sit inactive and build new wind farms with natural gas prices above $3.50. Since 2000, gas prices have rarely dipped below $3.50 and never for an extended period of time. To keep things in perspective, natural gas was about $2 per Mcf throughout most of the 1990s. Even if natural gas prices were to drop back down below $3.50 consistently (extremely unlikely), wind energy would still be less expensive than energy derived by constructing and permitting *new* gas-fired power plants.

"While natural gas prices spike without warning, the price of wind power has consistently decreased and will only get cheaper," said Colorado Public Interest Research Group's Stephanie Bonin. "In these tough economic times, reliability is important."[168]

Minimum Cost to America through 2025: **$4.2 Trillion**.

(More realistic cost to America through 2025: $5-7 Trillion)

MILITARY

An often overlooked cost of our current energy policy is the military. We spend a great deal of money every year to protect our oil interest in the Middle East. In fact, a recent report from the bi-partisan GAO (General Accounting Office), the watchdog arm of Congress, stated that the U.S. spends over $75 Billion per year in direct costs to maintain a military presence in the Middle East, and will spend at least $300 Billion on the current conflicts in Afghanistan and Iraq. These are just the direct costs. Indirect costs to the military (such as designing weapons systems specifically for desert climates, road and other infrastructure upgrades in the region performed by sub-contractors working for the Department of

---

[168] CO PIRG http://www.environmentcolorado.org/envcoenergy.asp?id2=9362

Defense which are not included in the direct costs, and other costs) have been estimated by some to be double that amount.

Clearly, it is impossible to place a cost on the lives lost so far in the Middle East – but obviously this cost in loss of life is inexcusable. Young men and boys are dying every day over there, and the Arab world is growing to hate us even more than they did before the conflict began. Although stability in the Persian Gulf is certainly desirable, and Saddam Hussein's regime was murderous and truly evil, it is obvious the cost in human life and greenbacks has turned out to be far higher than we ever anticipated. We are in this position because we have used up most of our oil and never got serious enough about creating affordable, cleaner fuels from renewable American-made sources such as hydrogen from wind and solar farms.

So, $75 Billion per year (I am assuming that this cost will not increase although it is inevitable that it will) x 20 years = $1.5 Trillion, plus $300 Billion for the 2003-2005 conflict is **$1.8 Trillion**. With The Freedom Plan, there would be little to zero need to have troops in the Middle East, so we could save the $1.8 Trillion cost. This military cost estimate also assumes that:

1. Another major conflict does not break out in the Middle East at any point in the next 20 years (unlikely unless we follow The Freedom Plan), and
2. Assumes that Saudi Arabia is able to forestall any major uprising among their large and growing ultra conservative Muslim factions who are fanatically anti-American. If the royal House of Saud fails in their attempts to maintain order and control of these radical Islamic factions, it is extremely likely that the U.S. will be asked (or decide to act on its own under the questionable pretext of national security) to quell the uprising and restore peace in this, the most oil-rich Muslim nation in the Middle East.

Documents recently declassified from the Eisenhower administration, available at the Eisenhower Library in Abilene, Kansas, state that the United States was prepared to be extraordinarily aggressive with military force to seize control of Middle Eastern oilfields.[169] In one meeting, a top U.S. official said access to oil was as important to the combined national security of the U.S., Great Britain and their other allies' as the proliferation of atomic weapons. Access to oil is as big of a deal as nuclear bombs? Well then let's solve that issue by eliminating the need for oil altogether! It appears the Eisenhower administration set a policy in 1958 on use of force to control Persian Gulf oil that would be adopted by most – if not

---

[169] Kansas City Star "50s Plan Prepared for Oilfield Seizure" by Steve Everly 1/31/04.

all – future administrations as well. In fact, according to a British intelligence document, the Nixon administration had a very similar contingency plan in 1973 as reported by *The Washington Post* on December 31, 2003. Nixon was a member of the National Security Council during the Eisenhower administration that originally adopted this stance of using military force to control oil in the Middle East.

According to Mr. Shibley Telhami, a senior fellow at the Brookings Institution in Washington D.C., author of *The Stakes: America and the Middle East* and a frequent contributor to CNN, CNBC, MSNBC, and other media outlets, the United States' influence over Saudi Arabia is waning. 18 percent of America's oil imports in 2003 came from the Saudis, who have the largest oil reserves by far in the world (262 billion barrels or 25 percent of the world's total remaining oil reserves; Iraq is in second place with 115 billion barrels, about 44 percent as much oil as Saudi Arabia).[170]

He predicts that over the next decade the United States may lose influence with Saudi Arabia – and there are signs that this is already happening. Mr. Telhami travels to the Middle East regularly, and said in an interview with the Kansas City Star[171] that "the Saudis' share of the global oil market will continue to increase, and they will have to be more responsive to their public than in the past...[which] puts a different set of pressures on them. ...They will have to be less sensitive to us...They have to be thinking strategically that they don't have to worry about U.S. policy as much – especially after Iraq."

He goes on to say that "the Saudis now have more clout over the U.S. because of Iraq. And there is the belief in the region that the Iraq campaign was a failure and the U.S. has less clout politically and militarily. The U.S. has considerably less leverage...The Saudis are by far the most important player, and their power is likely to increase because of their reserves. They also have the most surplus oil production. No other country has that, including Russia."

Mr. Everly, the reporter for the Kansas City Star in this article, pointed out that "the Truman, Eisenhower, and Nixon administrations all had contingency plans in case they lost control of Mideast oil fields. President Nixon, for instance, discussed using the military to seize the fields after the OPEC embargo. Do you think there is a possibility that such a policy is in effect now?" he asked Mr. Telhami. And the response: "I don't know that there is – none of us do...But I'm

---

[170] International Energy Annual and World Oil as reported in the K.C. Star on 7/23/04.
[171] "Is America Over A Barrel?" by Steve Everly of The Kansas City Star, July 23, 2004.

willing to bet you there were bureaucrats in the White House or State Department that were preparing policy papers...and a secret doctrine that involved oil. I think that when the archives are opened in 40 years, there will likely be some doctrine about the steps the U.S. would take to keep the oil flowing at acceptable prices."

Minimum Cost to America through 2025: **$1.8 Trillion**.

(More realistic cost to America through 2025: $2-3 Trillion)

ENERGY PRICE INCREASES

2004 was a year of record energy prices: gasoline hit an all-time high, diesel and other fuels from oil are at or near record high prices, natural gas prices have continued to hover around $5 per thousand cubic feet for the longest extended period in history. Even coal prices are rising to near record highs due to worldwide demand for coal from China, India and increased use of coal in the USA. The spot price for Central Appalachian coal averaged $41.50 per short ton for the week ending January 16, 2004, which was 33 percent higher than the same period in 2003.[172] As a result, some of the largest increases in electric bills were slapped on consumers and businesses in 2004 – in some cases these increases were as high as 18 percent in a single year. In fact, electricity costs in 2003 were already reaching 20-year highs, expected to rise by 4 percent assuming 'normal' weather to surpass nine cents per kilowatt hour on average.[173] With the latest increases in natural gas and coal prices, 2005 and beyond will likely set new record highs for electricity rates. And if we continue to see 'abnormal' weather, these prices could be significantly more painful, especially during hot summer months that are getting increasingly more sweltering. Federal Reserve Chairman Alan Greenspan called tight natural gas supplies (a key contributor to electricity prices) a "very serious problem" in testimony to Congress recently.

Those figures do not even include the utility bill spikes that many have seen solely as a result of the fuel cost component that some utilities are allowed to tack on to electric bills (already covered in the Natural Gas section of this chapter). In some cases these fuel price adjustment charges have doubled and tripled

---

[172] U.S. Department of Energy, as reported in "A Period of Prosperity" by Brad Foss of the Associated Press on January 26, 2004.
[173] USA Today, June 10 2003

consumer electric bills. The cost of energy is clearly increasing across the board, and in my opinion this is just the tip of the iceberg.

The United States has the lowest energy prices of any developed nation on Earth. That is a fact we can be proud of on one hand since it has helped lead to the most highly productive and efficient economy in the world. On the other hand, one of the reasons the U.S. has such low energy prices is that the true cost of all aspects of our energy is not paid for directly. The health damage, military costs, and environmental cleanup expenses of our existing energy policy are not yet reflected in the prices we pay at the pump and at the gas and electric meters.

This is likely to change in the near future, however. In Europe, fuel prices are two or times higher than ours here in the U.S. and have been for some time. However, many economies in the EU have government expenditures that are much higher as a percentage of GDP so higher overall taxes tend to overinflate fuel costs. In 2001 the electric utility industry in Europe implemented a groundbreaking mechanism that allows the true cost of power (including the effect that power plant emissions have on global warming) to be more accurately reflected in the cost of electricity. This mechanism utilizes carbon taxes that, in effect, penalize power producers that spew carbon dioxide into the atmosphere while burning fossil fuels. These carbon taxes increased the wholesale cost of producing power by burning fossil fuels. These added costs were largely passed on to consumers in the form of higher electricity prices, but these carbon taxes have also applied substantial pressure to utilities to switch to cleaner, less polluting sources of power on a wholesale level. Utilities that do not have the ability to generate clean, emission-free power due to a lack of wind, hydroelectric or solar resource can purchase carbon credits through an active carbon credit trading exchange that was created to facilitate a fair system of sharing the burden of switching to clean power. As a result, carbon taxes have cut CO_2 emissions in several member countries of the EU by nearly 5 percent in just three years.

It is my belief that carbon taxes are inevitable here in the U.S. Indeed, the Chicago Board of Trade began voluntarily trading carbon credits for the first time ever in December 2002. In both 2002 and 2003, the U.S. Senate passed legislation called a Renewable Portfolio Standard (RPS) as described in detail in Chapter 4. Although the bill never made it through the U.S. House of Representatives to become a law, this RPS would have required every utility in the nation to purchase 10 percent of their electricity from clean renewable sources by the year 2015. And, it allowed utilities in areas with poor access to wind, solar or other renewable resources to meet the requirement through the purchase of renewable energy certificates (RECs) or "Green Tags."

RECs are another trading mechanism similar to – and in effect the exact opposite of – carbon credits. RECs are bought, sold and traded at various trading desks across the country and around the world. More detailed explanations of RECs and how they can be used to facilitate the implementation of the Freedom Plan in Chapter 9.

Why is this important relative to renewable energy? As we learned in the previous chapters, clean, renewable energy can stabilize, and ultimately lower, energy prices. As a reminder, consider a wind farm. One of the wind turbine manufacturers (GE, which makes their wind turbines at a factory in Tehachapi, California) has recently announced that their 1.5 MW model – their top selling wind turbine – is about to be certified to last 50 years. Once a wind farm is built, it requires no fuel to operate – just lubrication of the components which can be accomplished using bio-based grease and oils made from corn or soybeans. It requires very little maintenance (once per month each unit must be checked for calibration and lubrication fluids but most of the rest of the work needed to maintain a modern wind turbine is performed by computers). And it will sit there, continuing to produce power whenever the wind blows.

So to make sure we all understand what this means, wind farms use no fuel, have minimal maintenance, and produce power for 50 years.

Does it make sense that the price of the electricity produced at a wind farm can remain stable? And would that give wind farms (and solar power) a clear advantage over fossil fuels? You bet it does.

The chart below shows a utility in Texas that purchases much of their power from outside suppliers (independent power producer, or IPP). One IPP sells them power from a wind farm. The price from that IPP is shown as the blue line in the chart below. Notice how the price of the wind energy remains absolutely stable roughly $30 per MWh over the entire year shown (2000). Compare that to the volatility of the electricity spot market and the price of natural gas.

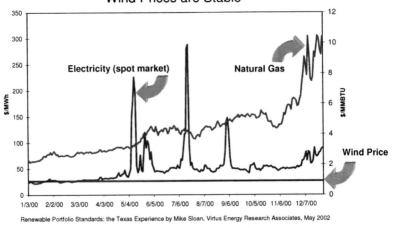

Renewable Portfolio Standards: the Texas Experience by Mike Sloan, Virtus Energy Research Associates, May 2002

Adding large quantities of wind energy generation can help stabilize energy prices, and will inevitably lower utility bills. Keep in mind that most states still have public regulatory agencies that monitor the actual cost of energy production. These regulators will eventually require the electric utilities in their state to lower prices as wholesale energy costs begin to decline after years of displacing fossil fuels with renewable power. Without downward pressure on the wholesale cost of energy production from wind power and other stable renewable sources, it is certain that utilities will continue to be granted new tariffs with higher energy prices and be allowed to add fuel adjustment charges to the bills to recover their costs of providing power to their customers. Even in states with deregulated electric industries, wind power can and should create a more competitive environment to stabilize and eventually lower energy costs.

The U.S. economy pays hundreds of billions of dollars each year towards power, light and heating bills. Average electric demand grows by 2-3 percent per year nationwide, and the average price we pay for electricity and heating increases regularly. If we make the conservative assumption that utility bills and fuel costs will increase at merely the rate of inflation, increased energy prices over 20 years (excluding fuel costs which were discussed earlier in this chapter) will cost the U.S. economy at least $2 Trillion over the next 20 years. But most experts contend that energy costs will continue to rise much faster than inflation. This should lead to another $2 Trillion of energy cost increases above and beyond the rate of inflation. Keep in mind that it is not just the raw energy commodity price increases that apply upward pressure to retail energy bills. Other factors can contribute, such as rising transportation costs to ship coal and other fossil fuels, more strict emissions standards being passed, legal costs from litigation with opponents of fossil fuels, and more. For example, Duane Richards, CEO of the Western Fuels Association Inc., based in Westminster, Colorado laments the rising cost of shipping coal by train cars: "The lowest rate I've heard is 25 percent increases, the highest is 100 percent," said. The association buys 20 million tons of coal a year for power plants owned by municipalities and cooperatives who are members of the association. Richards said it typically costs about $5 per ton to move coal about 500 miles. "Today I think the railroads are setting expectations for 500 miles that the price will be closer to $8 or $9 a ton," he said.[174] That is nearly a 100 percent increase in shipping costs, which is infuriating the utility industry since there are really no other realistic shipping alternatives.

That $2 Trillion figure merely reflects the expected increases of electricity demand and the impact of higher energy prices on utility bills. It does not include the impact or likelihood of drastic fossil fuel price spikes which could be caused by a) major supply disruptions, b) rapidly increasing worldwide demand for raw fossil fuels, or c) inclusion of carbon taxes and/or other factors that more accurately reflect the true cost of burning fossil fuels for our power.

Minimum Cost to America through 2025: **$2 Trillion.**

(More realistic cost to America through 2025: $4-6 Trillion)

ENVIRONMENTAL COSTS

The next major cost to U.S. taxpayers – and our economy – as a result of our current short-sighted energy policy is environmental clean up. It would be easy in this section for me to discuss the severe environment damage we are causing by burning fossil fuels as described in Chapter 3. But discussing it is about all I would feel comfortable doing, since any attempt to quantify this damage is tantamount to trying to place a dollar amount on the cost of lost lives resulting from our military defending U.S. oil interests in the Middle East. It is, in a word, impossible.

Therefore, for the purposes of this section – and to continue to strive towards being as conservative as possible – I will refrain from making any speculation as to the true nature of the damage we have done to Mother Nature. Instead, the focus shall remain strictly on the direct quantifiable cost of simply cleaning up the worst of damage we have caused.

---

[74] Denver Business Journal, "Coal Freight Costs Pinch Power Users" by Cathy Proctor, December 6, 2004.

Common sense tells us that burning fossil fuels wreaks havoc on air, water and land. But it takes hard facts to ascertain the actual direct costs of this damage. The indirect costs are quite difficult to accurately establish, so they will be largely omitted from this analysis.

The following examples should shed some light on how severe the damage is – and how large the tab for cleaning it up will be:

The Exxon Valdez oil spill ranks 22nd on the list of the worst oil spills in world history. Exxon estimates that they spent <u>$2.1 Billion</u> on the clean-up.[175] The true cost of such oil spills may never be known.

The cost of cleaning up the fuel additive MTBE damage to public water supplies is estimated to be at least <u>$29 Billion</u>.[176] Some new estimates hint that it may be several times that amount when the national impact of MTBE contamination is better known.

Petroleum product spills are more commonplace than many people realize. Oil, gas, diesel, propane, butane, naphtha, and natural gas spills and leaks occur on a regular basis in North America. For example, even in eco-conscious San Francisco, a pipeline that pumps petroleum from refineries in the bay area ruptured in 2004, gushing diesel fuel into a pristine bird marsh that serves as a key nesting ground for migratory birds. The 57,000 acre Suisun Marsh, about 25 miles northeast of San Francisco, is home to 700,000 birds. It was filled with 40,000 gallons of toxic diesel fuel. The cost to clean up this spill was not released to the public. These types of fuel spills occur more often than one might realize.

There are 1,233 toxic waste sites in America that are so polluted they are considered "Superfund" sites. These sites are located in nearly every state, are simply the very worst of tens of thousands of toxic waste dumps that will eventually need remediation as well. Cleaning each site can cost up to a billion dollars. Prior to 1995, the Superfund trust fund was supported by a tax on the chemical and oil industries. An environmental tax on corporations was also used to fund cleanup activities at sites where a responsible party could not be determined. The Superfund taxes expired in 1995, and Congress has not reauthorized them. In the meantime, EPA has drawn down the balance of the monies in the Superfund trust fund to cleanup sites. The trust fund ran out of

[175] http://www.evostc.state.ak.us/facts/qanda.html.
[176] Los Angeles Times article October 14, 2001, http://www.bugsatwork.com/044.ASP.

money in October 2003.[177] Since the Superfund tax expired, the trust fund has been reduced from a high of approximately $3.7 billion in fiscal year 1996, to $400 million in fiscal year 2002, according to the U.S. Environmental Protection Agency (EPA), Washington, D.C. Taxpayers now contribute approximately 53 percent of the revenue that is tapped for cleanups. Seven years ago, taxpayers provided only 18 percent of the funding.[178] While polluters may no longer have to pay to clean up the messes they leave in communities, the price tag on clean-ups has jumped dramatically: from $300 million per site in 1995 to more than a billion dollars this year -- a jump of more than 300 percent. And without sufficient funding, site cleanups are slowing down. This year an estimated 46 sites in 27 states will not be funded or will be inadequately funded.[179] Using a conservative estimate of $4 Billion per year (in today's dollars this is actually *less* than the $3.7 billion cleanup tab paid in 1996), cleaning just the existing Superfund sites will cost a minimum of <u>$80 Billion</u> over 20 years in today's dollars. Adding new sites to the list as they are discovered should push that price tag well over $100 Billion. Adding the costs of all the additional toxic waste dumps that today are not even included in this list could easily be multiple times that number.

A Superfund site

Photo courtesy The Sierra Club

The cost to clean up and dispose of radioactive waste is enormous. The U.S. Government Accounting Office says the cost to clean up just one site, the Paducah Kentucky uranium enrichment facility, will exceed $1.2 Billion by 2010.[180] For more than 50 years, the United States created a vast network of facilities for research and development, manufacture, and testing of nuclear

---

[177] GAO http://www.imakenews.com/flashpoint/e_article000181996.cfm.
[178] Waste Age April 1, 2003 Kim A. O'Connell Contributing Editor Arlington, Va.
[179] Sierra Club report September 30, 2004: http://www.sierraclub.org/toxics/superfund/.
[180] GAO report June 27, 2000.

weapons and materials. The result is that more than 7,000 sites at over 100 facilities across the nation have subsurface contamination, more than half of which contains metals or radionuclides and most including chlorinated hydrocarbons. DOE is responsible for remediating 2 trillion gallons of contaminated groundwater and 75 million cubic meters of soil and subsurface sediment. The groundwater volume is equal to about 4 times the U.S. daily water consumption, and the sediment volume would fill 17 professional sports stadiums. DOE estimates that, using current technology, cleanup will take 70 years at a cost of $300 billion.[181] And that is only the cleanup cost. Disposal costs are estimated to be billions more, including $12 billion recently announced in Congress for the next stage of preparing the Yucca Mountain nuclear waste repository. Hence, total costs could easily exceed <u>$400 billion</u> by 2025.

The cost of cleaning up the enormous quantities of toxic coal ash sludge at mining facilities (impoundments, as discussed in Chapter 3) and at power plants (in landfills) is likely to be hundreds of billions of dollars over the next 20 years.

Mercury, lead, arsenic, and other heavy metals that have leached from coal ash sludge, been dumped into our rivers and streams in the form of water discharge, and have seeped into public drinking water systems will cost untold billions across the nation. Retrofitting power plants to contain these poisonous chemicals after public demand finally forces power plant operators to add additional emissions controls will cost hundreds of billions more.

The more I have learned about what some companies have gotten away with, the angrier I get. It is reprehensible that the EPA does not even regulate mercury as a pollutant – one of the most poisonous, nastiest pollutants to be spewed out of every coal plant in the nation – as we learned in Chapter 3. Well that is only the tip of the iceberg. Many other toxins are not yet regulated, or the EPA has no jurisdiction over it, or news of the leak does not reach the right people, or any other number of factors may play are part in allowing environmental damage to continue unabated.

The good news is: things are slowly improving. Although the Bush Administration attempted to significantly weaken the 'teeth' in the Clean Air Act and hamper other progress made in the last decade, overall the enforcement level has generally increased since 1970. But these improvements come at a cost: better enforcement means a larger Environmental Protection Agency, more

---

government funds for Superfund site cleanup and countless other expenditures that U.S. taxpayers will have to incur to clean up other sites.

Furthermore, the cost to industry, businesses, state and local governments, and ultimately consumers and taxpayers for all of this environmental clean up work is shocking. Many estimates show that over the next 20 years this cost will be as high as $10 Trillion. The most common estimates range from $3 Trillion to $5 Trillion. Keep in mind that these estimates include only today's regulated and known toxins. As environmental testing continues to become more and more sophisticated and the public demands a higher and higher level of accountability and safety with respect to environmental damage, it is inevitable that new toxins will be identified, prioritized, and added to the list of cleanup projects.

Part of this cleanup cost goes to support environmental engineering and remediation firms which adds to the GDP and can create jobs. But the cost of cleaning up the damage after the fact is far, far higher than preventing it from ever occurring in the first place, so there is a tremendous amount of waste and unnecessary spending with this line of thinking. Nevertheless, in keeping with our conservative theme, we'll go with the lower estimate of $3 Trillion for environmental cleanup costs over 20 years.

Minimum Cost to America through 2025: **$3 Trillion**.

(More realistic cost to America through 2025: $6-10 Trillion)

HEALTH COSTS

As Americans, we are not well. We consume far too much sugar, do not exercise enough, drink carbonated beverages as if they were water (a typical soda can contains up to eight teaspoons of sugar and the carbonated water lowers the pH of our blood to acidic levels making us tired soon after the caffeine wears off), and eat too many refined foods packed with preservatives and chemicals. We are in very poor health as a nation, so it should not be surprising that we are the fattest nation on Earth and have soaring healthcare costs.

One of the most compelling reasons to adopt The Freedom Plan is the severe damage we are doing to ourselves from pollution related to energy. Our bodies simply cannot eliminate all the added toxins from air and water when we are already in a weakened state of health due to our hectic lifestyles and terrible eating habits. Men's Health magazine recently warned its readers to "cast a wary eye on these leviathans who have eaten their way to the top of the fishy food chain: shark, mackerel and swordfish. All have moderate to high levels of mercury." A study published in 2004 by the New England Journal of Medicine reports that "men with the highest levels of mercury in their blood had double the risk of suffering a heart attack."

Some of the more shocking health costs related to energy industry pollution and global warming are:

The healthcare cost to American children of environmental pollution (lead, mercury, childhood cancer, developmental disorders, neurobehavioral disorders, etc.) is estimated to be at least $55 Billion per year (nearly 3 percent of total U.S. healthcare costs)[182] or $1.1 Trillion by 2025 assuming no increase in annual overall medical costs.

In 1990, the annual cost of asthma to the U.S. economy was estimated to be $6.2 billion, with the majority of the expense attributed to medical care. A 1998 analysis using different methods estimated the cost of asthma in 1996 to be over $11 billion per year.[183] A leading, if not the primary, cause of asthma is air pollution. Assuming this cost has not risen since 1996 and will not increase at all over 20 years to be conservative, the total cost to our nation of asthma by 2025 is $220 Billion.

As the ozone hole becomes larger and the total ozone layer protection worldwide becomes thinner, more of the sun's dangerous ultraviolet radiation reaches the ground and our skin and eyes. Skin cancer in Americans grew by 6 percent per year in the 1970s, and in 2004 an estimated 55,000 Americans will contract melanoma (the deadliest of all skin cancers) and over 10,000 will die.[184] Skin cancer cases increased by up to 56 percent over a recent 10-year period in Europe and North America.[185] The ozone hole over Antarctica is now large enough to more than cover the entire United States (see image below).[186] Children in Australia are now required to wear sun bonnets or hats to school in

---

[182] The Center for Children's Health & the Environment & The Dept of Community & Preventive Medicine, Mount Sinai School of Medicine, New York, July 2002: http://www.ncbi.nlm.nih.gov/entrez/query.fcgi?cmd=Retrieve&db=PubMed&dopt=Abstract&list_uids=12117650.

[183] Dept of Health & Human Services Action Against Asthma, Strategic Plan, May 2000

[184] American Cancer Society.

[185] Natural Capitalism, by Paul Hawken, Amory Lovins, & L. Hunter Lovins, 1999.

[186] Planet Earth, German Aerospace Center, with Intro by Robert Hughes, 2002; Union of Concerned Scientists report "The Science of Stratospheric Ozone Depletion"

order to protect themselves from harmful UV radiation.[187] I have noticed along with many of my friends that it seems a lot easier to get sunburned than it ever used to be. The Center for Disease Control estimates that cancers cause $189 Billion in costs to the American economy. Skin cancer is one of the most common forms of cancer (nearly 1 million cases are diagnosed annually of all forms of skin cancer) but probably represents only about 10 percent of that cost, or $18 Billion per year (<u>$360 Billion</u> by 2025). On September 11, 2003, the region of ozone depleted air over the South Pole reached its second largest size in history at 10.9 million square miles, an area larger than North America.[188] The ozone hole has been carefully measured for decades; this recent measurement was surpassed only by the 2000 ozone hole of 11.5 million square miles, leading to increased fears of even more cases skin cancer.

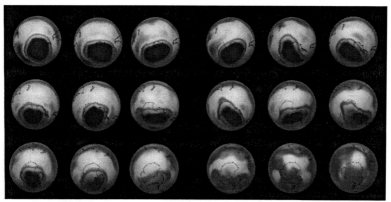

Ozone Hole Image courtesy of Planet Earth, German Aerospace Center

The total cost to society in 1993 of chronic obstructive pulmonary disease (COPD) was $24 Billion and respiratory cancer was $25 Billion.[189] Today those figures have more than tripled. Now that evidence shows power plant pollution and diesel fuel emissions exacerbate these and other ailments, what impact does our poor air quality have on these costs? Since the greatest source of air pollution – by far – is from power plants, the author estimates that at least one-third of these cases are linked at least indirectly to air pollution from burning fossil fuels. Further study on this issue is important to force utilities to clean up their act. That

---

[187] World Health Organization, Fact Sheet Number 261, July 2001.
[188] USA Today "Antarctica Ozone Hole Gets Larger" September 26, 2003.
[189] American College of Chest Physicians, 2000 "The Economic Burden of COPD" by Sean D. Sullivan, PhD; Scott D. Ramsey, MD, PhD and Todd A. Lee, PharmD.

would suggest a cost over 20 years of <u>$325 Billion,</u> assuming healthcare costs do not increase at all.

More than 100 brands of candy sold in California (and likely in other states), much of them made in Mexico, have tested positive for dangerous levels of lead.[190]   Lead is known to damage the nervous system, and is especially dangerous to children and pregnant women.   Coal-fired power plants are one of the leading sources of manmade lead pollution in North America.   Some studies say coal plants are now the leading source of lead pollution.   Mexico and other countries supply the U.S. with food rely heavily on coal for electricity production.

In the summer of 2004, MSNBC and other media outlets reported a groundbreaking study that was released linking power plant pollution to over 24,000 deaths per year.[191] Americans are dying prematurely due to energy industry pollution at a rate of at least 65 per day from asthma attacks, heart attacks, lung disease, and upper respiratory failure. It is impossible to assign a cost to so many of our neighbors and countrymen dying before their time.

At least eight (8) states are now suing power companies to force them to clean up air emissions.   The attorneys general for eight northeastern states and New York City filed a lawsuit in New York state court in 2004 to reduce power plant pollution.   They are trying to pressure five large power producers — American Electric Power Company, Southern Company, Xcel Energy Inc., Cinergy Corporation and the federal Tennessee Valley Authority — to clean up their emissions and help curb global warming.   The plaintiffs claim those power producers own 174 fossil fuel-burning power plants that produce 646 million tons of carbon dioxide annually, which is about 10 percent of the nation's total.   The attorneys general claim greenhouse gases like carbon dioxide could have catastrophic effects, including increased asthma and heat-related illness, depletion of drinking water supplies, a decline in fisheries and erosion of infrastructure.   Marc Violette, a spokesman for New York Attorney General Eliot Spitzer, declined to comment Tuesday on details but said the lawsuit would, "for the first time, put global warming on the litigation map. This is a precedent-setting, first-of-its-kind lawsuit," he said.[192] Clearly this lawsuit is the first of what could mimic the tobacco

---

[190] The Orange County Register, April 25, 2004.

[191] "Pollution is Blamed in Deaths" by Karen Dillon, Kansas City Star, June 10, 2004.

[192] CBS News http://www.cbsnews.com/stories/2004/07/21/tech/main630877.shtml.

industry litigation woes of the previous decade and cost the utility industry hundreds of billions of dollars.

Stray voltage, in-home electromagnetic radiation (EMF) and excess ground current has recently been discovered to cause significant health damage to humans – including childhood leukemia – and to farm animals. This problem has been made worse in recent years by the addition of numerous electronic components, especially computers, to the electric power system (the grid). These electronics tend to amplify the levels of toxic radio wave pollution from power lines in the walls. Recent settlements and lawsuit victories in Wisconsin and other states have shown that schools and buildings where these electric current problems are addressed have seen dramatic drops in a matter of days in the rate of asthma, fatigue, headaches, blood sugar levels in diabetics, and more. Other symptoms of radio wave sickness[193] are: difficulty concentrating (children are commonly treated with the over-prescribed drug Ritalin when potentially misdiagnosed as having Attention Deficit Disorder), dizziness, nausea, headaches, dizziness, nausea, difficulty concentrating, memory loss, irritability, depression, anxiety, insomnia, fatigue, weakness, tremors, muscle spasms, numbness, tingling, altered reflexes, muscle and joint paint, leg/foot pain, "Flulike" symptoms, fever. More severe reactions can include seizures, paralysis, psychosis and stroke. Cardiac: palpitations, arrhythmias, pain or pressure in the chest, low or high blood pressure, slow or fast heart rate, shortness of breath. Respiratory: sinusitis, bronchitis, pneumonia, asthma. Dermatological: skin rash, itching, burning, facial flushing. Ophthalmologic: pain or burning in the eyes, pressure in/behind the eyes, deteriorating vision, floaters, cataracts. Others: digestive problems; abdominal pain; enlarged thyroid, testicular/ovarian pain; dryness of lips, tongue, mouth, eyes; great thirst; dehydration; nosebleeds; internal bleeding; altered sugar metabolism; immune abnormalities; redistribution of metals within the body; hair loss; pain in the teeth; deteriorating fillings; impaired sense of smell; ringing in the ears. This relatively new trend could become one of the most litigated areas of this decade, pitting the healthcare industry squarely against the energy industry.

Entire neighborhoods from a town in Ohio had to relocate their families recently due to efforts by a regional utility to curb pollution flowing from an 80 story tall smokestack. The thick black smoke was traveling

---

[193] The National Foundation for Alternative Medicine "The Health Effects of Electrical Pollution"

hundreds of miles to towns that were threatening legal action, so the utility modified its emissions control systems. But the change displaced dozens of families in Chesire, Ohio who were suddenly choked out by a bluish cloud of sulfuric acid that was now flowing into their neighborhoods, causing layers of soot to settle on everything in their homes and burning the delicate tissues in their noses and lungs. The solution? The plant's owner bought the town and tore it down after evacuating all its citizens.[194] Sometimes the solutions chosen by utilities can cost even more than anticipated for emissions controls. These occasional added costs, the total extent of which are unknown today, may double or triple emissions reductions efforts at some fossil fuel plants.

Overall healthcare costs have been soaring. Certainly the role that increased toxicity is playing cannot be overlooked. How much energy industry pollution adds to the escalating costs of healthcare remains to be seen, but it is not reasonable to presume that it is a significant contributor?

Minimum Cost to America through 2025: **$4 Trillion.**

(More realistic cost to America through 2025: $10-20 Trillion)

Total Minimum Cost to America through 2025:

$20.5 Trillion

(More realistic total savings through 2025: **$35-61 Trillion)**

These are frustrating calculations, for sure. Well, surprisingly, there are numerous additional costs of our current energy policy that I have intentionally omitted from this exercise. Again, it is important to remain conservative in order to deflate the potential objections of critics who are certain to resist the Freedom Plan for whatever reason. Demonstrating that we can easily launch volley after

---

[194] USA Today "Pollution Unites Town, but Solution Tears it Apart" March 16, 2004.

volley of abundant economic evidence to support the Freedom Plan will hopefully stifle much of that resistance.

Just to hint at some of the excluded costs of our myopic energy policy, I have listed a few of these potential additional costs below. As you can see, many of these costs could be significant.

OTHER COSTS NOT INCLUDED:

The Homeland Security department is spending tens of billions of dollars to supposedly make America safer, capitalizing upon the 9-11 tragedy, by instituting an Orwellian-inspired security program slated to cost taxpayers $177 billion dollars for fiscal years 2004-2005[195] and could easily cost U.S. taxpayers another $3.5 Trillion over 20 years. If we were not so hopelessly dependent on Middle Eastern oil, we would not have to send our cash and troops to that region, which would not enrage the Islamic militants nearly as much, and would reduce or perhaps eliminate the need for an entire Cabinet position and department dedicated to Homeland Security.

The book Natural Capitalism, published in 1999, points out that weather-related claims are putting staggering burdens on insurance companies: "in 1998, violent weather caused $90 Billion of damage, which represents more weather-related losses than all those of the entire decade of the 1980s," and the authors warn that "the world faces danger of being torn apart by regional conflicts instigated in part by shortages or imbalances in resources such as oil, water, fish or minerals." We're already seeing ample evidence of that prediction. For example, the hurricane season of 2004 has been the worst ever, with damage in south Florida alone estimated to exceed $10-20 Billion from hurricanes Charley, Francis, and Ivan. This is just a fraction of the weather related damage we will see over coming years. As ocean levels rise and affect low-lying coastal regions, as temperatures get warmer causing mudslides in California, permafrost melting in Alaska, and who knows what else around the rest of the country, these costs could be crippling over the next 20 years. Wildfires in California cost more than $2 Billion in 2003.[196] These costs are conservatively estimated to be $40 Billion per year, or $800 Billion over 20 years.

Legal costs of defending lawsuits brought by victims of health damage from energy industry pollution (the movie "Erin Brockovich" was based on such a real

---

[195] Government Accounting Office, September 2004; Left Turn Magazine Issue #13.
[196] U.S. News and World Report Special Edition "The Future of the Earth: A Planet Challenged from the Arctic to the Amazon" July 2004, p 13.

case), litigation and other legal action as it relates to the enforcement of emissions standards and cleaning up pollution could easily soar into the hundreds of billions of dollars in coming years.

$500 Million per year to the Black Lung Association (20 more years of that cost leads to another <u>$10 Billion</u> cost to the USA).

Higher travel industry costs, as jet fuel for the aviation industry and diesel fuel used by cruise lines continues to rise. Airlines, cruise lines, rental car agencies, and other travel industry companies are sure to pass many of these costs on to their customers.

U.S. taxpayers have already spent over $100 billion for our nuclear subsidies of the past. Utilities are obligated to pay only a fraction of the $100 billion cost of disposing of their high-level nuclear waste; these costs will be mostly paid for by tax- and ratepayers. Bailing out the nuclear utilities will cost taxpayers another <u>$300 billion</u>, and the industry plans to sue if we do not. Despite these enormous taxpayer subsidies, nuclear power is still by far the most expensive type of electricity. Electricity ratepayers are paying not only for nuclear power's higher rates, but also for nuclear power plant cost overruns, premature decommissioning and financial disasters like New York City's Shoreham nuclear plant fiasco. Finally, the nuclear industry has profoundly insufficient accident insurance.[197]

Total government subsidies and tax credits to the fossil fuel industry are estimated to be at least $30 Billion per year (20-year cost of <u>$600 Billion</u>).

Acid rain damages paint on vehicles, bridges, and other structures. Automakers now use acid-resistant paints at a cost of $61 million per year for just the cars and trucks sold in the U.S.[198] 20-year cost of <u>$1.2 Billion</u>.

For every degree of global warming in the Midwest (average temperatures have risen by one degree in just the past 15 years), corn crop yields decline by 10 percent.[199] Since 90 percent of corn is eaten not by American consumers but by livestock (poultry, cattle, and hogs),[200] an increase in food costs across the board could result from a decrease in corn supply. Due to the energy-intensive nature of agriculture, wholesale and retail food costs are expected to rise significantly

---

[197] Mendocino Environmental Center, 2001.

[198] Clear the Air, Washington DC.

[199] According to ecologist Christopher B. Field of the Carnegie Institution.

[200] US Department of Agriculture.

faster than inflation over coming years. This could add to a tally of hundreds of billions of dollars over the next two decades.

The opportunity cost of lost benefits to our economy if another nation were to lead the world in developing the technology and intellectual resources to lead the world towards a massive conversion to clean power. This shift to clean power will represent the largest shift of wealth that world has ever seen. The United States should not let herself lose this one. If we do not act swiftly, Europe and Asia, who are embracing this massive shift to clean power, will be celebrating their wildly successful economic rewards at the expense of the United States in 20 years.

What is the cost to the economy of running continued enormous trade deficits, led primarily by our imports of fossil fuels?

A lack of national pride; since we must rely on others to meet our energy needs. We became a net energy importer for the first time ever in the 1970s. Although it is difficult to quantify this cost, it undoubtedly affects the morale of the nation, which could affect productivity. How do we measure the cost of losing our pride as other regions of the world prosper while we suffer for a while until we wake up and get serious about clean power?

Foreign Policy: how much is spent on smoothing ruffled feathers or buying loyalty from oil-rich nations under the banner of the State Department? These amounts which include much more than direct foreign aid are surprisingly significant, reaching into the tens of billions of dollars per year.

In May 2004, consumer groups reported that consumers in America have been hit by domestic oil companies with $250 billion in price increases since 2000.[201] A lack of domestic oil company competition from mergers has led to huge profits for the oil companies and increased prices to consumers that would not have occurred with more competition. If that trend were to continue through 2025, it would reasonable to assume that the total additional cost to consumers of this lack of competition would equal at least another $1.25 Trillion over 20 years.

A recent study determined that a sustained $10 per barrel rise in oil will reduce global output by 0.5 percent after one year.[202] In just two years, average oil prices have jumped $35 per barrel but the sustainable increase will likely be at least $25

---

[201] Report at www.consumersunion.org or www.consumerfed.org/oilprofits.pdf.

[202] The Economist, "A Crude Awakening" joint study by IEA, OECD, and IMF p. 12.

per barrel. That would suggest a 1.25 percent decrease in global output. What impact will that restricted world growth have on U.S. GDP and exports? One might conclude that this one factor alone could cost the U.S. economy hundreds of billions of dollars per year, or trillions over 20 years. Therefore, this argument proves once more that converting to clean power and eliminating dependence on oil – while drastically reducing greenhouse gases – will actually improve the economy.

Firewood costs are rising as more people turn to trees for heating fuel in America as heating oil and natural gas costs soar. In New England, firewood is up 37% in the winter of 2004-5 compared to the prior year, and prices for cords of firewood in the West have also jumped by as much as 30% in some areas. Rising fossil fuel costs can affect our pocketbooks in a myriad of ways that one may not predict until it begins to actually pinch the budget.

By some estimates, we lose 27,000 jobs for every billion dollars of additional oil imports.[203] In 2004 when average annual oil imports grew from $80 Billion to over $230 billion, what was the true economic cost to America?

None of these extra costs, which could easily exceed **$5 Trillion** in additional costs by 2025, are included because a) it is not necessary since the case has been made convincingly without them and b) it is, quite frankly, rather disturbing when one adds it all up. We have been foolish as a nation to look the other way while politicians in Washington D.C. flounder in their attempts to reduce dependence on fossil fuels and foreign oil, but we only have ourselves to blame for not requiring an effective solution. We must begin to <u>demand</u> drastic change. The Freedom Plan requires immediate and massive action, and is the only solution available – the <u>only</u> solution – that guarantees success since it does not rely on the government. I would be willing offer the profit from an entire wind turbine for one full year to anyone who can present another solution that can guarantee the successful conversion of America to clean power within 10 years without relying on government support or intervention.

CONCLUSION

If we continue doing what we're doing as a nation, we will have to pick up the tab for an additional $20 Trillion (at least) over the next 20 years. That's $20 Trillion <u>more</u> than what we are currently expecting to have to absorb.

---

[203] "Set America Free" a Blueprint for U.S. Energy Security, 9/27/2004.

When we implement The Freedom Plan, at a total cost of $2 Trillion, we will not have to pay that $20 Trillion. Hence, we will save $18 Trillion over 20 years, or more realistically, we may save closer to $40 Trillion when all the additional costs are accounted for. And, with dependence – finally – on inexhaustible renewable energy, once the infrastructure of The Freedom Plan is in place, we should begin to see a steady and significant decline in overall energy prices. *That* is something few of us have been brave enough to realistically wish for. But it can become a reality with enough widespread endorsement of The Freedom Plan.

In addition, the economic impact of The Freedom Plan should easily exceed $6 trillion by 2025. In addition to the many examples of positive economic stimuli already listed throughout this book as a result of shifting to clean power (millions of new jobs, additional tax revenues, wind energy royalties to landowners, etc.), an improved energy system can make our nation more efficient. For example, a modernized electricity system (power grid and related systems) will reduce the cost to the economy of power quality problems and "enable productivity improvement and GDP growth."[204] The EPRI Roadmap estimates that modernizing the electricity infrastructure could increase economic productivity by 0.7 percent per year over business-as-usual conditions. In the United States alone, this translates into about $3 trillion *per year* in additional GDP by 2025. That suggests an economic impact of **$45 trillion** over a 15-year period after the grid has been completely updated by year five of The Freedom Plan. Once again, the figures of this book are meant to be conservative, but it is easy to see how powerful the positive impact could be to the U.S. economy when we embrace clean power without reservation.

Total Minimum SAVINGS to America through 2025:

$20 Trillion +

(More realistic total savings through 2025: **$40-66 Trillion**)

Total NET Savings to America (after paying for The Freedom Plan):

---

[204] Electric Power Research Institute (EPRI) "Electricity Technology Roadmap 2003.

$18 Trillion +

$18 Trillion of total savings represents a significant savings to society. With 100 million families in America, this represents a gross savings of $9,000 per year for 20 consecutive years for every family in America. **That's a gross savings to society of <u>$750 per month</u> for every family in America** – which is far greater than the average total expenditures on energy per family today! And that does not even include the positive economic impact of at least $6 Trillion that is to be expected. Nor does it include many of the other likely costs of our current path. And finally, it does it include the savings we will enjoy from lower energy costs expected to begin appearing after 50 percent of The Freedom Plan is completed. These savings will be reflected in our energy bills, fuel bills, food costs, and nearly everything else since energy permeates nearly every facet of our lives.

In other words, if we do not launch the Freedom Plan in the very near future, somehow every family in America will have to pick up the tab for our current energy policy to the tune of $750 per month, every month, for the next 20 years. And that is being conservative! The total cost could be triple that amount or higher and someone is going to have to pay for it.

So we are now faced with a choice. 2005 will go down as an historic year where the United States reached a crossroads. As a nation, we will make a choice either proactively through public awareness and demanding collective action by a majority of our citizens, or by default with inaction. Thus, here are the choices:

Continue doing what we're doing (take no action) and by default we will have made the choice to pay an additional $20 Trillion or more on energy over the next 20 years. That price tag will have to be absorbed by one of the following options:

    a. **Dramatically higher energy bills and fuel prices, increasing our average monthly costs by well over $750 per month,** OR

    b. **Experience much higher taxes to allow the government to pay the tab.**

Keep in mind that in 2004, the total U.S. government budget – <u>all</u> expenditures – was about $2 Trillion and much of that goes to entitlements such as Social Security, Medicare, Medicaid and interest on the national debt. That leaves only a few hundred billion dollars for roads and bridges, defense, national security, housing, education, parks, law enforcement, research and development, existing energy funding and all other federal programs. Finding $1 Trillion per year (that's half of total U.S. government spending!) in the current already strained federal budget is impossible. There is absolutely <u>no way</u> the government can pay this tab for us without huge tax increases that would exceed an average of $9,000 per year per family in additional taxes.

Make the proactive decision to implement the Freedom Plan. The U.S. government will not make this choice for us. Governments are, by nature, reactive and not proactive. As much as we may want our government (or any government) to be visionary and take leadership on this issue, there are simply too many vested interests influencing not only the President but also the real lawmakers: Congress. It is unlikely that we will ever elect an individual who is powerful enough to sell this vision to Congress and the nation.

Note that the cost of the entire Freedom Plan is only $2 Trillion. That seems like a large number, but it is the same amount that the U.S. government spends in total every year. It is not that far out of reach! All it will take is a sustained commitment on the part of a multitude of individuals, scattered and fragmented non-profits who all have equally altruistic missions but who could be so much more powerful if all were aligned towards one common objective such as The Freedom Plan, and support from hundreds of companies and entrepreneurs who join the Clean Power Revolution and offer their time, energy, and creativity. This collection of stakeholders can be immensely powerful. This group is all that is necessary to successfully implement The Freedom Plan. It truly can become a reality with enough broad support and if sufficient numbers of people can relinquish their ego enough to embrace a plan that they themselves did not create.

Therefore it is up to us, as citizens, to demand the launch of The Freedom Plan. Only through grass roots marketing and a groundswell of support for the Freedom Plan or another feasible alternative will this overhaul of our energy policy – and therefore the protection of our health and environment – take place.

The next chapter will demonstrate why this is such a no-brainer and you'll learn that many others agree that the time has come for action on this issue. You'll be surprised how simple it can be to actually make a difference. One person <u>can</u> make a difference: you are a peaceful army of one. I encourage you to consider the consequences – and the cost you'll ultimately have to pay – if you do not take action.

Chapter 8:

"DUH!"

As my younger brother Nathan used to say, "Duh..."

In other words, isn't this conversion to Clean Power a 'No-Brainer'? Of course it is! Then why haven't we done something about it?

To review...

...if making electricity causes so much health damage,
...since our current energy policy is poisoning the air, water, wildlife, and environment around us,
...if we import ~70 percent of our oil today with plans to import 85 percent by 2025,
...if the cost of fossil fuels continues to remain high,
...if it is true we have more violent hurricanes and weather than ever before,
...if we will have to pay an extra $1 Trillion per year of added costs,
...and we have enough wind potential to economically power our entire grid plus make enough hydrogen to power every car and truck in America...

...THEN SHOULDN'T WE DO SOMETHING ABOUT IT? Well, duh, yes of course we should. And we will. We must. But who or what company (or group of companies) will lead the way? And when?

As a nation, we have not yet taken charge of our future to convert to clean, renewable, cost-effective, *American-made* power. Let us explore why, and determine what our chances are of implementing the Freedom Plan or another suitable alternative. To quote the famous line from the "Lord of the Rings" movie trilogy distributed by 20th Century Fox studios: "there is always hope." believe in a strong, independent America. I know there are millions of other who feel the same way. We have an opportunity in front of us now that only comes along once in a lifetime – to show the world how creative we can be i solving our biggest problems. Today, the cost – and reliable supply – of energy, the lifeblood of any economy, is our most pressing issue. The good news is, a cost-effective solution exists. And, I am absolutely convinced there is enough grass roots support to make this vision a reality.

DO PEOPLE REALLY WANT CLEAN POWER?

Strong public support is absolutely critical for a successful conversion to clean power on a scale as massive as that suggested by The Freedom Plan. This chapter will offer surprising results from studies on this topic that will hopefully both enlighten and entertain you.

The three current events topics that received the most media coverage in 2004 are, in no particular order:

1. The Iraq war,
2. The Presidential election, and
3. The impact of fossil fuels and foreign oil on our lives.

If you think about it, all three topics are closely related. The primary point of contention in the 2004 Presidential race is the debate over Iraq, linking topics #1 and #2. No other topic received more press coverage in the race than U.S. policy in the Middle East.

A close parallel can be drawn between the Iraq war and issue #3: our dependence on foreign oil from the Middle East (where 65 percent of known oil reserves exist). Common sense, which is actually not very common at all, tells us that a significant underlying reason we have a military presence in Iraq is to defend our national interest in the oil reserves of that country. Other reasons we attempt to rationalize the unacceptably high cost in both dollars and human life are: to attempt to 'stabilize' the region, and to create a new artificially induced ally to U.S. interests. Sure, possible Iraqi links to terrorist cells and the risk of Saddam Hussein possessing or attempt to create WMD were other reasons to be concerned. But in my opinion, we went to war to show the world that the U.S., Great Britain and a few of our key allies were more than willing to use our military strength to secure future access to the lifeblood of our economy: oil.

Hence the Iraq war, the 2004 Presidential race, and the challenges of the Fossil Fuel Age are inextricably related. People in America are far more intelligent than the media gives us credit for. We acknowledge our shortfalls as a nation, including our helpless dependence on foreign oil. Although most people do not realize how bad the energy situation really is, people all across the country (and the world) intuitively know that we have a problem. We know that our efforts in the Middle East are directly or indirectly related to oil. And we recognize that oil will eventually run out or get too expensive to extract (again, that is common sense). But what you may not realize is that a surprising number of people would like us to do something about it.

Over 1,000 surveys, polls, and studies have been conducted over the past several years on the consumer and business demand for clean power in the United States. These studies have been performed by very credible organizations, such as Stanford University, Platts Consulting, the Department of Energy's National Renewable Energy Laboratory (NREL) and Pacific Northwest Laboratory, the Edison Electric Institute, the Electric Power Research Institution (this organization is actually funded by investor-owned electric utility companies), the Environmental Protection Agency, The Public Interest Research Group, Time magazine, countless state government energy offices, non-profit organizations, and media outlets. They all say pretty much the same thing: that at least 80 percent of American consumers want more wind and solar power injected into the electric grid, and that 47 percent of all businesses want more clean power sources and less dependence on foreign energy.

According to one study, 64 percent of Americans would try harder to be green if more environmentally friendly products and services were readily available to them or if they thought their efforts would make a difference. Those earning more than $50,000 per year are more likely to pay more for clean power than other income levels, as are households with three or more people. However, Americans in the lower income brackets would be more likely to go green if they knew it would really help. Ignorance about how to make environmentally friendly choices is a primary reason why more Americans are not making the choice to go green.[205]

And, many of these polls studied whether consumers would pay a little more for their power if it was clean and renewable. One of the largest studies, conducted by the Edison Electric Institute, found that 22 percent of Americans would be willing to pay an extra $21 every month, on top of their current electric bill, if their power was guaranteed to be clean and renewable. This study assumed the customer received no benefit other than the heartfelt satisfaction of "doing the right thing" by doing their part to support clean power. That study, indicative of most other studies, indicates that nearly 75 Million Americans (22 percent) are

---

[205] Media Central Inc., a PRIMEDIA Company, Sep 2002 "Being Green" by Rebecca Gardyn.

willing to purchase clean power for an extra $21. Yet as of December 2004, only about 1,000,000 Americans had enrolled on a Green Energy program. That means less than 0.4 percent of Americans have actually enrolled, when at least 24 percent are willing to make such a purchase if they knew it was an option. In other words, at least 74 million Americans have already agreed to support clean power with their pocketbooks but just do not yet know how they can do their part.

Therein lies the problem: only about 20 percent of utilities across the nation even offer a Clean Power option to their customers, and many of these have pitifully small marketing budgets to promote these programs. Therefore, few consumers know they even have a choice between the dirty, polluting, toxic power they are now buying versus clean, renewable power. It is encouraging to note, however, that the growth of these programs has been robust: the number of kWh of green power sold from 1999-2003 grew by 57 percent per year among just the energy companies offering such an option[206] and by 71 percent per year overall (including third party marketing companies that bypass utility companies to sell Green Power directly to consumers in the form of Green Tags, which are explained in the next chapter). This makes Green Power marketing one of the fastest growing segments in the enormous energy industry.

But it is easy to grow quickly when the Green Power industry is so small. With only 1,000,000 customers on clean power in this country, I have to say that is an embarrassing result. Especially considering 75 Million would make the purchase if they just knew it was available. To make matters even more frustrating, the European Union has done a far superior job of marketing Green Power and Green Tags than the U.S. Although modern-day wind energy technology was originally perfected in California in the 1980s, and the United States led the world in total kilowatt hours produced from the wind for nearly 20 years, the U.S. wind energy industry was surpassed by Germany in 1998 and now even Spain leads the U.S. As a whole, the European Union has installed about eight times as much wind energy generation as the entire United States. And, the citizens of the EU support clean power in droves.

For example, in many European countries, nearly 30 percent of consumers have already enrolled on a Green Power program, paying more for their power to make it clean and renewable. That represents tens of millions of consumers in Europe choosing clean power for a small premium, versus an embarrassing one

---

[206] Group Veritas Report 2004 "The Mainstreaming of Green Power"

million here in America. The results of this strong consumer demand in Europe are astonishing.

Germany, which is by far the largest economy in Europe, has had a very strong Green Energy marketing and education initiative for years. In the last 10 years, Germany has invested over $30 Billion to build new wind farms, plus considerable efforts have been made with solar, geothermal, landfill gas and other renewable sources. Germany's nearly 20,000 MW[207] of wind energy capacity is enough to power 8 Million U.S. homes (equivalent to all of New York City) or about 15 Million European homes (European homes are nearly twice as energy efficient as the U.S. homes). About 12 percent of the country's electricity now comes from wind energy, with some states in Germany powered over 30 percent by the wind. This industrious nation, which is about the size of the state of Florida, now has three (3) times more wind farms than the entire United States!

Denmark has built so many wind farms that at times the wind blows strong enough to power 100 percent of the country! On average, Denmark receives 25-40 percent of their energy from the wind[208], leading the world in per capita wind energy production. The largest wind turbine manufacturer in the world is Vestas (which recently merged with NEG Micon) and is based in Denmark, along with other top manufacturers of wind turbines. Denmark has approved an initiative to rapidly attain 50 percent wind-powered status, with plans to increase that to 100 percent by approximately 2040.

Spain, a small but passionate country roughly half the size of Germany and about the size of an average Midwestern state, recently surpassed the United States as the world's second-largest (#2) producer of wind-generated electricity. A major manufacturer of wind turbines (Gamesa) is based in Spain, and GE has a manufacturing plant there to support the strong European demand for wind power.

Great Britain recently passed legislation requiring the electric industry to generate 20 percent of their power from renewable energy by the year 2017, one of the most aggressive timetables set forth by any country. Most of that energy will come from wind power. This effort is motivated in part due to the steady decline of oil production from the North Sea.

---

[207] European Wind Energy Association.
[208] Danish Wind Energy Association.

One study, performed by the Electric Power Research Institute (EPRI)[209], segmented the entire population of the United States into six distinct groups:

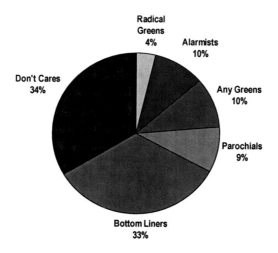

Pie Chart courtesy of Electric Power Research Institute

Traditional green power marketers view their customers as **"Radical Greens"** and **"Alarmists"** consisting of 14 percent, or almost 45 Million, of all Americans. These two groups will buy green power at a premium for the social benefit and because it is the right thing to do.

"Any Greens" and **"Parochials"** represent 19 percent of the market or nearly 60 Million Americans and may be persuaded to buy green power for a premium if there are additional benefits for themselves or the local community, or if it supports a worthy cause.

"Bottom Liners" (33 percent) may be persuaded to buy green power, but only if the price is the same or lower than traditional power or if there is a financial incentive involved (such as a customer referral program or a business opportunity), and **"Don't Cares"** (34 percent) typically won't buy green power for any price.

---

[209] EPRI Green Power Guidelines, Volume 1.

It is the opinion of the author that at least 10 percent of the population is just nuts (i.e. crazy) so it does not matter what you are doing, they simply won't respond. I think most of those nuts would fall in the "Don't Care" category.

The Radical Greens, Alarmists and Any Greens represent 24 percent (75 Million) Americans and will be the 'low-hanging fruit' for any initiative that attempts to capture the customers willing to support clean, renewable power. The Freedom Plan can be successful with only one-tenth (10 percent) of this 'Green' group enrolling on a clean power program with a participating Freedom Plan clean power vendor, or *wildly* successful with 90 percent or more of this group choosing clean power. But the Freedom Plan is actually designed to acquire at least 50 percent of the 'Green' group plus a modest 10 percent of the remaining three market segments. Assuming this goal is attained, the Freedom Plan will not only be successful, it will be completely implemented in just 10 years. All it takes to be free of the shackles of foreign oil dependence, with zero air pollution, zero water pollution and zero health damage from burning fossil fuels, is enough consumer demand and a sound plan of action. It sounds simple, and it is, but it will not be easy.

Another interesting study was performed by Platts Consulting of Denver, Colorado. This report shows how many people would likely choose a Green Energy program – and pay a little more for it – on a state by state basis. The average shows that about 24 percent of the population is willing to pay more for clean power, but there are large discrepancies by state. The states with the highest demand for renewables (in alphabetical order) are Connecticut, Hawaii, Massachusetts, and New Jersey while the lowest demand is expected to be in Arkansas, West Virginia, and Wyoming.[210]

---

[210] Platts Consulting , 2003.

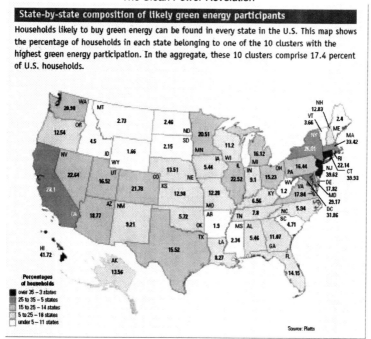

Image courtesy Platts Consulting

These studies assume that the buyer of the Green Energy gets only one thing: satisfaction. This satisfaction results from that fact they are: a) consciously choosing clean, renewable power; b) doing their part to reduce dependence on foreign oil (through lower use of #2 diesel and other fuel oils commonly used in electric power generators); c) lowering toxic pollution and emissions; d) indirectly supporting the construction of renewable energy resources; and e) doing the right thing.

Therefore, the conclusion one might draw from analyzing these two studies is that the U.S. marketplace is wide open with excellent growth potential in many states and good growth potential overall, having only 1,000,000 current Green Energy customers of 24 Million who simply want the satisfaction of doing the right thing. It appears that the retail Green Energy industry is ripe for rapid expansion, one into which many entrepreneurs, individuals, non-profits and companies will enter and prosper.

However, as you will learn in the next chapter, The Freedom Plan is sponsored by at least one company that offers a unique clean power program that offers multiple benefits, in addition to simply wanting to do the right thing. These other benefits include an optional charitable tax deduction and an optional monthly gift certificate program which combined lower the net cost of the clean power premium. Spend around $100 for some simple energy efficiency gains in your home and you can be net positive (actually saving more money than you spend) by choosing clean power and energy efficiency. And, your participation helps build new wind turbines in your country or region. This allows The Freedom Plan – and the Green Tag program that will drive it – to become more palatable to the non-Green market (the Bottom-Liners, Don't Cares and Parochials). Consequently, this expands the clean power market significantly and increases the likelihood of The Freedom Plan's ultimate success. There are many companies and organizations who have officially endorsed The Freedom Plan, providing multiple choices for consumers to help further ensure the success of its realization.

MEDIA COVERAGE

As mentioned earlier, the impacts of the Fossil Fuel Age (health damage, environmental damage, increasing costs, being dangerously dependent on energy from outside our borders, etc.) are one of the three most covered topics in the media in 2004. However, a cursory review of mainstream media publications, radio shows, and television news clearly indicates that reducing dependence on foreign oil and fossil fuels has become not only one of the most-covered topics but that *clean power is actually the most popular topic in current events today*.

For example, nearly all the major news magazines in the U.S. had a recent cover story that hint at how popular the Clean Power message has become. Some recent examples follow:

National Geographic: In the September 2004 issues there are 68 pages of photographs and stories, covering the many dangerous existing impacts of global warming, and dire predictions of future impacts of global warming as a result of fossil fuel dependence based on sophisticated computer models. On page 75, this issue eerily warns us that global warming will cause more intense hurricanes – just weeks before the worst hurricane season in history.

USA Today: there were many weeks in the second half of 2004 when a Clean Power issue (cost of oil or natural gas, health or environmental damage of fossil fuels, oil supply fears, economic woes caused by these issues, etc.) was a major (or featured) story in USA Today. In the Money section, a lead story about Clean Power and/or the cost of fossil fuels was on the cover every single day for an entire week in 2004.

The Wall Street Journal, The New York Times, the news wire services including the Associated Press, the Washington Post, the Chicago Tribune, and countless other publications had lead or feature articles in 2003 and 2004 about the myriad of problems caused by the Fossil Fuel Age. A majority of these articles mentioned wind power, solar power or hydrogen as a part of the solution – often reminding us that we have done a pathetic job of weaning ourselves off of fossil fuels and foreign energy and that it is time to get serious about it.

Yet even with all this media coverage, my colleagues and I have been greatly disappointed that neither candidate in the 2004 Presidential race really addressed the issue, let alone made it a major part of their platform.

HOLLYWOOD GETS INTO CLEAN POWER

In the summer of 2004, the blockbuster film "The Day After Tomorrow" was released by 21st Century FOX. This movie was a huge summer hit, raking in nearly $190 Million in the U.S. and over $500 Million worldwide. The plot is laced with its fair share of Hollywood drama, but the science behind it is sound. The story is about a climatologist in Great Britain who discovers an anomaly in the global ocean conveyor system (essentially, the moving conveyor belt that moves warm waters from the Indian Ocean and equatorial regions to the north Atlantic and provides Europe and North America our temperate climate). This anomaly is caused by global warming, since ice melting in glaciers around the world causes freshwater runoff to pour into the ocean faster than ever before. The heavier saltwater sinks lower and the salinity levels in the ocean are modified. The conveyor relies on warm southern water releasing its heat in the northern hemisphere, then sinking to lower depths and moving back to the south to allow the warmer, lighter water coming from the south to flow over the top of the heavier, cooler water. But when light freshwater is added in ever increasing quantities to the equation, the freshwater competes with the warm saltwater moving to the north, interfering with the ability of the conveyor to continue moving. A rapid shut down of the conveyor ensues, resulting in cataclysmic changes in weather patterns, climate, and rainfall – and concludes with an ice age covering the northern half of the United States and Europe.

The movie's timeline, undoubtedly to add suspense and drama to the film, shows the entire process taking place over the course of a few days. In reality, this theory is sound and is based on real science, but would likely occur over the course of months or years, not days or weeks. However, it may come as a surprise to learn that some very conservative, non-partisan, and reputable institutions have determined that such a scenario is plausible, albeit on a slower timeline. One such institution is the very conservative United States Department of Defense, warning us in a recent Pentagon report that "global warming could rival terrorism as the greatest threat to National Security."[211]

Consider that statement for a moment. Throughout history, mankind has fought brutal and bloody wars in attempts to pursue primarily one or two agendas: religious conversion or control of the inherent natural resources available in a region (water supply, arable farmland or grazing land, precious metals or gems, and more recently wars over minerals, coal and oil). For centuries, the natural resources controlled by a nation of people were their primary measure of wealth.

If we see a major shift in climate, or even a modest shift, the most arable farmlands today could become less productive or even non-productive. Every one degree increase in global temperature results in a 10 percent reduction in corn production in the Midwest.[212] You may not think that matters if you do not eat corn on a regular basis. But you actually eat a lot more corn than you realize: only about 3 percent of all corn produced in the United States is eaten by humans[213]. The rest is eaten by livestock (cattle, chickens, turkeys, and hogs), meaning unless you are a vegetarian you are indirectly eating a lot of corn in the form of hamburger, steak, pork chops, bacon, chicken, turkey, etc. The global temperature has already risen by more than one degree Fahrenheit in the last 50 years, and is expected to increase another 1º F in the next 20 years[214] although no one really knows how fast the temperature will rise.

Assuming this is true in the Midwest, would it not be safe to say that there will be impacts elsewhere in the world? If farming output declines by even 10 percent in some populous countries like India and China (both have nuclear weapons at their disposal), what kind of social impact would events such as widespread famine, severe drought, and more violent weather have on the already tenuous military ceasefire agreements around the world?

Hollywood, as usual, has capitalized on these potential fears with a well-made film about the risks of global warming. It should continue to have an impact around the world as people purchase the DVD, rent it, and watch it on premium movie channels.

---

[211] Knight Ridder Newspapers "New Weather Patterns Could Stir Turmoil" by Seth Borenstein, February 29, 2004, report available on www.ems.org/climate.

[212] National Geographic "Global Warning"

[213] United States Department of Agriculture at www.usda.gov.

[214] National Geographic "Global Warning"

But it does not stop there. Many actors have embraced the Clean Power Revolution by their actions in addition to their acting: the number of moviestars who have purchased hybrid cars is growing rapidly (including Cameron Diaz, Matthew Broderick, Billy Crystal, and others). A Los Angeles limousine company, Evo Limo, recently retrofitted a number of their vehicles to run on compressed natural gas instead of gasoline.[215] Woody Harrelson, Brad Pitt, Cameron Diaz, and Kevin Richardson (of the Backstreet Boys) are clients, among others. This lowers emissions by over 80 percent (cutting smog in southern California) and does not contribute to foreign oil dependence (natural gas comes primarily from the United States and nearby areas such as the Caribbean and Canada). As a result, the demand to use these limos is so high that many celebrities have not only begun using this service, some have decided it is the *only* limo service they will use.

Dennis Weaver, Gwenyth Paltrow, Darryl Hannah, Edward Norton, Ed Begley Jr., and Tom Hanks are other celebrities living and promoting a green lifestyle by relying on clean energy. They either use their celebrity to promote green programs, purchase hybrid vehicles or renewable energy systems for their homes, donate their time or funds to clean energy causes, or a combination of these. The Environmental Media Association gives awards each year to television programs, films, and documentaries promoting clean power and sustainable practices.

INVESTORS CHOOSING CLEAN POWER

There has been a significant movement over the last few years in the investment community called SRI (Socially Responsible Investing). There is strong demand by the investing public to put their money into companies that are defined as being socially responsible. This includes, of course, companies involved in renewable energy and hydrogen.

Franklin Templeton Funds, which now owns a significant stake in wind farms in the U.S.

Warren Buffet (financing a $320 Million wind farm through a subsidiary of Berkshire Hathaway)

Merrill Lynch, American Century, Bank of America and Vanguard all have SRI mutual funds

---

[5] Stefan Lovgren in Los Angeles for National Geographic News April 19, 2004.

The Association of SRI has seen double-digit growth in membership and conference attendance every year for over 5 years

Watchdog groups, such as the Investor Responsibility Research Center, are holding accountable such large firms with poor sustainability practices as GE, GM, Exxon Mobil, TXU and others. They are applying pressure to these and other large firms to force them to provide better disclosure of carbon dioxide emissions and other pollution, as well as attempting to get these firms to use and support renewable energy and be socially and environmentally responsible.[216]

Shell Oil Company, one of the largest companies in the world, now boasts to its investors that it has invested over $1 Billion in hydrogen and another $1 Billion in wind farms in the last few years. The following quote, taken from a speech given by the CEO of Shell Renewables, Karen DeSegundo, was printed recently in several reports issued by the company: **"By the year 2050, 1/3 to 1/2 of all of the world's energy will come from renewable sources."** This coming from an oil company and one of the largest companies in the world, at that. The pressure from the public to convert to clean power, whether from consumers buying products or consumers making investments, is being heard loud and clear; even by those in the oil industry who take the time to listen.

A Growth Industry

Renewable energy is one of the world's great growth businesses. Power from sources such as the sun, wind, and fuel cells has grown from a $9.5 billion global industry in 2002 to $12.9 billion in 2003, according to research consultancy Clean Edge, which projects the market to reach $92 billion worldwide by 2013.[217] "These are no longer alternative technologies," says Ron Pernick, co-founder of Clean Edge. "We are at the tipping point." Venture capitalists are now jumping into clean power as well: in the first six months of 2004, $76 million in VC investments were pumped into alternative energy, almost double the 2003 figure, says a PricewaterhouseCoopers' MoneyTree Survey.

Venture capital firm Draper Fisher Jurvetson has invested over $20 million into clean power startups since 2002. "We are seeing the same ramp-up as we've seen with the Web," says Draper Managing Director Raj Atluri. General Electric acquired Enron's wind-power business in 2002. And it entered the solar energy business in April, 2004, by purchasing AstroPower, the nation's largest solar

---

[216] N.Y. Times "Report Faults Big Companies on Climate" by Barnaby Feder 7/10/03.
[217] BusinessWeek "Racing to Energy's Great Green Future" October 8, 2004.

equipment supplier. Both companies have been growing at a double-digit annual rate, while GE's total sales rose only 1 percent last year.

CONCLUSION

The energy industry is the largest industry in the world by far.

It is being forced to shift to clean, renewable sources of power, and the pace of that shift is steadily increasing.

Over 240 Million Americans (80 percent) say they <u>want</u> clean, renewable power – and nearly 75 Million (24 percent) are willing to pay <u>more</u> for it.

In the U.S. today, only 0.2 percent of our homes are enrolled on a Green Tag program, compared to roughly 30 percent in many European countries.

The United States has the world's best wind energy potential – by far.

Yet America generates less than 2 percent of our power from the wind.

It is time to get serious about leading the world in the Clean Power Revolution. Europe is kicking our tail in regards to wind power development. As a proud American, I do not want to be left behind.

One suggestion, although certainly not the only solution, is to use grass roots marketing to guarantee that The Freedom Plan is implemented. Such a solution does not rely on getting any laws or regulations passed by any government; it needs zero government spending to be successful; it does not need permission from any electric utility or energy company; and can begin immediately. To that end, on Earth Day 2004, I launched a company called Krystal Planet with help from my wife, Alysia, and a group of like-minded colleagues who are passionate about converting our nation to clean power.

I am now certain that Krystal Planet will successfully implement the Freedom Plan, because in just a few short months after the launch of the concept we attracted thousands and thousands of people who have enrolled on FutureWindSM, a clean power option. This is the first and most critical step of The Freedom Plan. With the continued success of Krystal Planet and other organizations that officially endorse The Freedom Plan, combined with the widespread distribution of this book, The Freedom Plan is inevitable. Hence, these are very exciting times to be an American and have the opportunity to be actively involved in the Clean Power Revolution.

I am humbled by the popularity of this book and its principles, and am grateful to you, the reader, for taking the time to learn about these important issues. Clearly, this Clean Power Revolution is so much larger than I am, and much larger than any one company such as Krystal Planet. Therefore, please accept my sincere apologies for promoting Krystal Planet, a company in which I have an interest, along with other clean power companies in the next chapter. My intentions are certainly not to offend anyone. Whether or not you choose to take action to support this cause, and whether or not you take that action through Krystal Planet is your business. However, The Freedom Plan will never get off the ground unless a well thought-out marketing plan is pioneered and aggressively executed by creative, entrepreneurial companies. In fact, there are a host of non-profit organizations, businesses and government entities who have endorsed The Freedom Plan. If you wish to do your part to support this cause, I encourage you to take action – through any company that guarantees a potent contribution to The Freedom Plan.

If you, the reader, discover an alternative, equally elegant method to further this cause, I implore you to vigorously support it with all your available resources. Otherwise, I hope you will find the following chapter about the status of the Clean Power Revolution to be interesting, informative, and perhaps even entertaining.

Chapter 9:

PLANET SAVERS

"When our resources become scarce, we fight over them.
In managing our resources and in sustainable development, we plant the seeds of peace."

WANGARI MAATHAI of Kenya, winner of the 2004 Nobel Peace Prize

The world cannot stop global warming without the participation of the United States. The Freedom Plan is a market driven solution that will substantially improve the U.S. economy over the next two decades. For it to succeed, the most critical component is getting enough consumers and businesses to choose clean power instead of dirty power. That single concept will do more to jump start The Freedom Plan than any other initiative, any government policy or tax credit, or anything else.

Once I realized this fact in 1999, I knew that developing wind projects the old-fashioned way (by getting a Power Purchase Agreement from a utility that is often cool to the idea) was not the fastest way to convert the world to clean power. Therefore, Krystal Planet was formed to show the world how to choose clean power using grass roots marketing – and to help guarantee success for The Freedom Plan. As the message about clean power continues to spread, many other companies and organizations have begun to contribute in their own way to the Clean Power Revolution. However, a fragmented approach is not as powerful as a unified front. Therefore, this chapter provides a glimpse into some of the companies who are leading the charge in the Clean Power Revolution. This chapter describes the organizations who are embracing The Freedom Plan and how their efforts are saving the planet.

HOW KRYSTAL PLANET SUPPORTS THE FREEDOM PLAN

Krystal Planet, formed in 2002, sells a product called a Green Tag to consumers all over the world. This company is just one example of an organization supporting The Freedom Plan. Green Tags are also known as a Renewable Energy Credits, Renewable Energy Certificates, Tradable Renewable Certificates, and in Europe, they are called Renewable Obligation Certificates. The common term Green Tag encompasses all those formal names. The Krystal Planet Green Tag product, called FutureWindSM, sells for $30 per month and represents a written guarantee that 1,000 kilowatt hours (kWh) of clean, renewable wind energy are injected in the transmission grid. A $10 Green Tag called Krystal WindSM includes 400 kWh of clean wind energy, and is more appropriate for apartments, condos, smaller homes, and most dwellings outside of the United States where average energy consumption is much lower.

The concept is really quite simple: spread the word about clean power being a choice anyone can make using grass roots marketing, and enroll as many customers as possible. Every few thousand customers that enrolls on FutureWindSM allows Krystal Planet's parent company (Krystal Energy Corporation) to build another 1.5 megawatt, $1.6 Million wind turbine every year. These wind turbines last up to 50 years, and every 1,000 kWh they generate from the wind means that a power plant in the region does not have to burn dirty coal or gas to create those same 1,000 kilowatt hours of energy.

WHAT IS A GREEN POWER CERTIFICATE

Green Power Certificates, or Green Tags, were originally developed in the early 1990s in California by the owners of wind energy facilities and their utility customers to track the 'green' attributes of the power. They found there was a premium value for clean power over fossil-fuel 'dirty' power. The electric grid cannot distinguish between a 'green' electron from a wind turbine and a 'brown' electron from a fossil fuel power plant. Nor is it possible to send electrons through the grid to any particular home or business, since they flow to the nearest load, such as a light bulb or air conditioner motor. The 'green' electrons get mixed up with all the other electrons in the power pool. Therefore, a tracking mechanism was needed to accurately account for every 'green' electron injected into the grid. Wind farm owners meter the electrons injected into the grid and unbundle the 'green attributes' of that renewable power and sell it separately in the form of Green Tags. So if you want clean power coming to your home you buy generic electrons from your local utility, and then you must buy your Green Tags separately. Some utilities offer a green premium program which provide your Green Tags on your power bill, but most (about 80 percent) utilities only offer dirty power. Whether or not your own utility gives a choice to buy Green Tags, you can always choose to purchase them from a third party marketing firm such as Krystal Planet or any number of others. Often, the Green Tag products differ in number of kilowatt hours, whether the power comes from an existing or soon-to-be-built renewable facility, special features (such as tax deductibility, free gift, customer referral program), certification method, and type of green power generation (wind and solar having the highest value, followed by new and existing hydroelectric dams, biomass, landfill gas, and finally, some Green Tag marketers unfortunately consider burning natural gas as a qualifying green source of power).

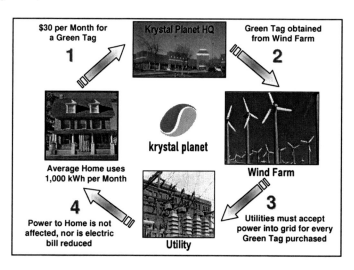

So, a Green Tag represents the "green-ness" of the power. It puts extra revenue into the hands of those who take the risk to build these multi-million dollar wind facilities. Because energy from wind farms and other renewables costs a little more today than burning fossil fuels at old power plants that are not required to install modern pollution controls, Green Tags indirectly support the construction of more renewable projects. They also represent pollution reduction.

If you live in a home or apartment, regardless of whether you own or rent, and you want to be able to say that your residence is powered 100 percent by clean, renewable, wind power, you have only two choices:

1. Install a windmill in your back yard on a 50 foot pole. It will cost $20,000-35,000, the wind is probably not strong enough or consistent enough, and your neighbors and zoning board are probably not going to be too happy about a 50 foot tower. That's not a good option. The only other choice you have today is...

2. Choose clean power by purchasing Green Tags. You buy your power (electricity) from one source (your local utility) and you buy Green Tags separately (either from your utility, in the rare case that they offer a green program, or from a Green Tag marketing firm). You will pay an extra amount each month to "Green Up" your home. This does not reduce your electric bill in any way.

There are over 20 companies in the U.S. marketing a wide variety of Green Tags, including Krystal Planet. Plus, roughly 20 percent of all U.S. electric utilities offer a Green Tag program to their customers. Although Green Tags transcend state and national borders, most of these green power marketing firms operate regionally, a few operate nationally, and even fewer (less than five) operate internationally. Krystal Planet's Green Tag product, called FutureWind℠, contains 1,000 kWh of Green Tags, 100 percent of which come from wind. Those 1,000 kWh produced by a conventional power plant create pollution such as mercury, sulfur dioxide, nitrogen oxide, carbon dioxide, dangerous solid particulate matter, lead, cadmium, trace uranium, arsenic and carbon monoxide. The United States Environmental Protection Agency (EPA) – working with the Department of Energy – created emissions reduction standards for Green Tags (discussed in detail below).

So, FutureWind℠ customers actually receive five distinct benefits:

1. A written guarantee that 1,000 kWh of 'green' electrons will be injected in the transmission grid (the average U.S. home uses slightly less than 1,000 kWh each month).
2. A direct, measurable reduction in pollution since those 1,000 kWhs are not created by burning fossil fuels.
3. The ability to claim that your home is pollution-free.
4. The ability to claim that your home is powered by the wind.
5. The comfort knowing that the purchase contributes to building a new wind turbine (see escrow account below).

Green Tags come in two flavors: those from existing wind turbines, and futures Green Tags which contractually provide their owner the exclusive rights to Green Tags produced by a wind turbine that has not yet been built. Although Green Tags are not securities, a futures Green Tag is similar to an oil futures contract, which represents a barrel of oil that has not yet been pumped from the ground. Several Green Tag marketing companies sell futures Green Tags, which directly contribute to the construction of new wind turbines. Krystal Planet utilizes a separate escrow account for FutureWind[SM], depositing a portion of the $30 monthly cost (at least the $5-10 wholesale amount of the green tag) into an escrow account used exclusively to build wind turbines. Once the escrow account contains $600,000, a new $1.6 million wind turbine is built (the cash from the escrow account allows the remaining balance of $1 million to be financed). Then, the next $600,000 deposited into the account builds a second wind turbine, and so on. Krystal Planet's objective is to enroll 1 million customers on FutureWind[SM]. Surprisingly, it takes only about 5,000 customers on FutureWind[SM] to deposit $600,000 into escrow in a year. In other words, every 5,000 customers buy enough Green Tags to build a new 1.5 MW wind turbine – every year. In ten years, those 5,000 customers will have added 15 MW worth of wind energy generation – enough to power over 8,000 homes for 50 years.

Another benefit of futures Green Tags is their purchase can allow wind turbines to be constructed in any area of the world. New wind generation can be built In the United States even if the local utility in the area has no interest in willingly buying the power output of that wind turbine. In other words, buying futures Green Tags can, in effect, force utilities to clean up their power generation. How is this possible? Laws passed after the 1970s oil embargoes, called PURPA[218] (for Public Utility Regulatory Policies Act) established what is known as a Qualifying Facility (QF). Simply put, these laws provided the legal framework under which independent power producers could build power plants and connect them to the national grid without permission from the local utility. Once a power plant is granted QF status, the utility serving the area where it is built is required by PURPA laws to allow access to their transmission grid <u>and</u> pay the owner of the QF for every kWh they inject into the grid. This payment is provided at "avoided fuel cost" – which is essentially the going wholesale rate for power in the area. Remarkably, any new renewable generation facility, including a wind energy facility, is *automatically* granted QF status. Consequently, with enough customers buying futures Green Tags, new wind turbines can be constructed even in those areas where utilities are hostile to renewable energy! This notion that utilities are invincible to changing their old dirty ways is nonsense, thanks to PURPA and futures Green Tags such as FutureWind[SM].

As we learned in the previous chapter, 75 Million Americans are willing to pay extra for clean power. Yet today, the largest Green Tag marketer (Green Mountain Power[219]) has just 600,000 customers nationwide. The total number of consumers paying for green power in the United States is estimated to be about 1

---

[18] See www.ferc.gov.

[9] Green Mountain Power of Austin, Texas. Be cautious of their claims, however: see www.boycottgreenmountain.org.

million – far less than the 75 million who say they would do so if given the choice. And many estimates in Europe tally nearly 50 million people are paying more for their power to make it clean and renewable. Buying a green tag is akin to buying premium gas at the pump. Premium gas costs more, but it usually contains 10 percent ethanol (which is made from corn, a renewable resource), it burns cleaner (less pollution), provides better gas mileage, and it is better for your vehicle's engine. Buying premium power for your home via Green Tags also costs more, but it comes from a 100 percent renewable resource, is cleaner power than the conventional power you are buying now, and it is better for your *body's* engine since it creates none of toxic pollutants that poison our air, water, and food.

GREEN TAG CERTIFICATION

To add a measure of credibility to this relatively new industry, most Green Tag marketing firms employ a third party certification process. There are several standards offered by different certification bodies. Not all Green Tags are certified (in fact, many are not). There are at least twenty companies in the USA selling Green Tags of all different varieties. Nearly all Green Tag certification methods audit the following characteristics:

They ensure that for every kWh of Green Tags sold, a kWh of clean power is injected into the grid.

They validate Green Tag content: when a customer buys a Green Tag comprised of 100 percent wind, there is no biomass, landfill gas, natural gas, or other non-wind resource included.

They certify that the same Green Tag is never sold twice by requiring the marketer to use a unique serial number for each Green Tag sold.

One of the most popular certification standards, though not necessarily the best, is provided by the Center for Resource Solutions (CRS): green-e. CRS is a small non-profit that relies on funding from public and private donors, including the EPA and Department of Energy. But CRS does not have enough funding or personnel resources at the time of this writing to certify futures Green Tags. Other certification bodies include the Climate Neutral Network, ERCOT (Electric Reliability Council of Texas) for Texas Green Tag marketers and others.

Because Krystal Planet utilizes futures Green Tags – and because FutureWinds™ is marketed through independent business owners instead of easily controlled internal sales representatives – CRS is not an option for certification. Hence, FutureWind[SM] is not a green-e certified product and probably never will be.

Instead, Krystal Planet's Green Tag products employ a far superior certification standard. Krystal Planet is the only company in the world to use a large public accounting firm to certify its Green Tags. Krystal Planet currently utilizes a $2 Billion-per-year public accounting firm with 585 offices in 114 countries to annually audit every Green Tag sold by Krystal Planet – and audits every dollar deposited into the escrow account. This adds tremendous credibility and reassurance to all FutureWind[SM] and Krystal Wind[SM] customers: they know that their dollars are going towards validated, certified 100 percent wind green tags and that not a single dollar of the escrow account goes to anything other than building new wind turbines.

A business could do the same thing. For example, Sprint Corporation, based in Overland Park, Kansas, a suburb of Kansas City, chose in April 2004 to purchase 2.5 Million kilowatt hours of Green Tags on a 2-year contract. This purchase enables Sprint to be able to claim that one of the buildings on the campus of their world headquarters is now 100 percent powered by clean, renewable wind energy and is the first totally pollution-free building in the Kansas City area. Yet these Green Tags comes from a wind farm hundreds of miles away from the Sprint building.

GREEN TAG ENVIRONMENTAL IMPACT

For example, every household on FutureWind[SM] (or an equivalent Green Tag product from other clean power companies) for a year eliminates the following emissions:

172 grams of mercury: just 1 gram of mercury dropped into a 25 acre lake is enough to poison all the fish in that lake. Mercury is incredibly toxic, linked to autism, Alzheimer's and heart disease. Power plants are responsible for 61 percent of all airborne mercury in the U.S., a whopping 5,000 grams every minute of every day. Each home on FutureWind[SM] for one year saves 4,000 acres of lakes from becoming polluted.

67 pounds of nitrogen oxide which creates smog, air pollution, dangerous ground level ozone, all of which causes headaches and exacerbates asthma. Cases of asthma in children in the U.S. have tripled in the last 2 decades, and asthma is the leading cause of missed school days in the Northeast.

55 pounds of sulfur dioxide, which causes acid rain, eye damage, and poisons entire ecosystems.

12 tons of carbon dioxide, the leading global warming gas. This is also equivalent to planting 36 acres of carbon-dioxide absorbing forests.

Hundreds of pounds of solid particulate matter, which helps induce asthma attacks and contributes to upper respiratory illness, heart attacks and lung disease.

Hundreds of milligrams of lead, cadmium, arsenic, other heavy metals, and trace uranium, as well as dozens of pounds of poisonous carbon monoxide.

Clearly, one person **can** make a big difference in cleaning up our air, water, and energy supply.

FUTUREWIND℠ ADVANTAGE

All Green Tags purchased – from any source – will indirectly support the implementation of The Freedom Plan. For that reason, I wholeheartedly support any effort to sell Green Tags by any company – and I support any credible Green Tag product. In order to guarantee success of The Freedom Plan however, the Krystal Planet team has developed a Green Tag that appeals to both environmentalists <u>and</u> mainstream America. Without broad appeal, Green Tags will not become ubiquitous and consequently The Freedom Plan may not become a reality. Krystal Planet's FutureWind℠ product has five unique competitive advantages:

1. **100 percent wind**: many Green Tag products include content from incinerators, which burn biomass products. This process, while renewable and cleaner than coal, emits substantial amounts of carbon dioxide and other pollution. FutureWind℠ contains 100 percent pure, clean wind energy.
2. FutureWind℠ is a **future Green Tag**, directly contributing to the development and construction of new wind turbines (in many cases, these wind projects will be constructed locally in the state, region or country where enough customers are enrolled).
3. It includes an optional **customer referral program** – similar to MCI's Friends and Family™ and Verizon's "IN" Network™, Krystal Planet offers certain benefits to customers who spread the word about clean power and refer other FutureWind℠ customers.

4. It can be 100 percent **tax deductible** if you choose to donate your Green Tag to a 501c3 non-profit. This significantly lowers the net cost of FutureWind℠ to make it more economically attractive (check with your own tax advisor).

5. It includes an **optional monthly thank you gift** – receive a gift certificate every month worth up to $15 (participating vendors offer $5-15 in cash value for every $5 that Krystal Planet provides to the vendor on behalf of the customer) as a way of saying thank you for doing the right thing. The value of the tax deduction and the gift combined can lower the FutureWind℠ net cost significantly. Those who decline the gift increase the dollars going into the escrow account, which helps build new wind turbines more quickly.

Other Green Tag marketers have interesting and attractive Green Tag products as well. I encourage you to study the benefits and differences of each, and determine whether they have officially endorsed The Freedom Plan before selecting a clean power service. In closing, I encourage you to buy clean power for your residence in the form of Green Tags, or take the time to research this concept further if necessary. Whether you buy FutureWind℠ from Krystal Planet or you buy Green Tags from another source, please *do your part* to support The Freedom Plan and buy Green Tags somewhere.

GRASS ROOTS MARKETING

The Clean Power Revolution will be most successful if the companies involved utilize powerful grass roots marketing and word-of-mouth advertising to build their collective customer base. As an example, although dozens of utilities and third party Green Tag marketers spend millions on advertising, Krystal Planet spends almost nothing on advertising. Instead, many customers are acquired by paying a referral bonus to existing FutureWind℠ customers. Akin to recommending a good movie or restaurants to people you are in contact with, this concept counts on customers recommending FutureWind℠ to their friends, neighbors, and co-workers. And, a financial incentive is provided to do so. Many other companies successfully employ such strategies. Examples are:

Verizon Wireless™ – Verizon's IN™ network allows customers who refer other customers to Verizon wireless to call them for free (Verizon phone to another Verizon phone), providing a financial incentive to help add more Verizon customers.

United Airlines recently offered their customers a bonus of 10,000 frequent flyer miles if they simply Refer Four Friends to the program.

PayPal™ (an eBay™ company) pays their customers a $100 bonus to refer someone that opens a new merchant account with PayPal™.

- The MCI Friends and Family ™ program allowed customers to make long distance calls for free or reduced rates to people they encouraged to sign up on the MCI long distance service, again providing a financial incentive to help MCI build their network of customers.
- WOW Cable Company – the Wide Open West cable TV company offers one full free months' cable service to any customer who encourages someone to switch to WOW.

These types of customer referral programs encourage customers to be the best source of advertising.

Krystal Planet utilizes a worldwide sales force of independent dealers. Other companies operate on a local or regional basis with employee or contractor sales representatives. The independent marketing professionals and clean power advocates primarily operate from home with very low overhead. They are called Energy Consultants (or Ecos) at Krystal Planet and can open their own worldwide clean power business for less than the cost of a good pair of shoes. These Ecos market FutureWind^{SM} Green Tags to consumers worldwide in exchange for ongoing monthly commissions on the sales they make. Ecos also sell energy efficiency products (including the Green Machine which improves fuel economy in any vehicle by 10-20 percent) and perform Home Energy Reviews where they test the energy efficiency of homes and recommend products to reduce energy bills by an average of 20-40 percent per month. Homeowners can even choose to purchase renewable energy products for their home, such as solar shingles, small residential wind turbines, and home hydrogen systems (with a built-in electrolyzer) to power the entire home and up to two vehicles with clean, safe, pollution-free hydrogen. Modest part-time effort allows many Ecos to earn enough income to become Energy Independent. This means they earn more than enough to pay for all their family energy costs (electricity, natural gas heating, and fuel for their vehicles). Many Ecos choose to reinvest a portion of their earnings into energy efficiency products and renewable energy systems for their homes. Some Ecos working full-time earn thousands or tens of thousands of dollars per month selling Green Tags, renewable energy systems, and energy efficiency products. Therefore, Krystal Planet is actively engaged in implementing two critical elements of The Freedom Plan: energy efficiency and establishing a market for small, distributed hydrogen fuel cell power plants and hydrogen production at home from wind, sun, and water. Other companies are also making steady progress in the Clean Power Revolution. I encourage you to research which companies are supporting The Freedom Plan and find out how you can help.

Chapter 10:

MAKING A DIFFERENCE

"Become an army of one."
Anonymous

It is amazing what a difference one person can make. The air we breathe and the water we drink in the United States are so toxic and polluted that we are poisoning ourselves as we learned in Chapter 3. And the worst culprit of that pollution – by far – is electricity generation. My wife Alysia and I have become exasperated by the American political system. No matter how hard we work at the state, local, and federal level to influence politicians to do something about this terrible energy policy we live under, we cannot seem to make a dent in the political armor of the energy industrial complex. Both political parties have failed miserably in weaning America from foreign oil – and now we are becoming dependent on foreign natural gas. The energy industry is so large that contributions to politicians from environmental groups simply do not make much difference. The most powerful way to force a change, in fact to guarantee change, is for you and your contacts to buy Green Tags for your home and business from a company that endorses The Freedom Plan. Krystal Planet has already begun leading the way in this Clean Power Revolution. Many other companies are joining Krystal Planet by endorsing The Freedom Plan. With your help promoting this book about The Freedom Plan, we cannot fail in our quest to convert America to clean renewable power that we can make forever right here inside our own borders.

"By simply starting a task, you have already completed 50% of it!"

To get the clean power train moving down the proverbial tracks, getting it moving is much more difficult than keeping it moving once it already has some momentum. Hence the need to kick-start The Freedom Plan to get it going so it can gain momentum on its own. Fortunately, Krystal Planet has already acquired thousands of Green Tag customers and will continue to be successful. And many other companies and organizations are now endorsing The Freedom Plan officially to add more momentum to the cause. I am hopeful that this book will help educate tens of thousands more about the possible solution to our energy crisis that is now before us. It appears the message is slowly beginning to spread and make an impact. According to the American Wind Energy Association, voluntary green power programs (selling Green Tags) are helping to bring new wind farms online. 1,200 megawatts (over $1.2 Billion of new investment into wind farms and other renewable energy projects) have been built since 1994 as a direct result of Green Tag programs.[220] *That is enough to power over 600,000 American homes using market demand alone.*

This chapter will demonstrate how much difference your own single monthly Green Tag contribution makes to help implement The Freedom Plan – and clean up our air and water. Because of people like you, the reader, I have been thoroughly impressed by how a small group of people can accomplish so much in such a short period of time when the cause is just. For example, one company supporting The Freedom Plan, Krystal Planet, was officially launched on international Earth Day in 2004 (April 27, 2004). By the time of the first convention on the weekend of November 5th, 2004, the company's revenues had increased by an average of 40 percent per month in 2004 and it had already acquired over 2,500 residential customers in 48 states and 15 countries (see chart below).

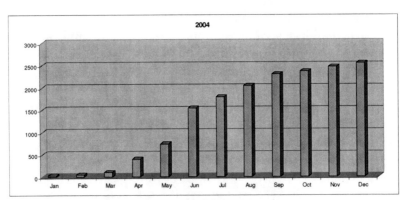

Krystal Planet FutureWind℠ customers

With 2,500 FutureWind℠ customers already on the books, this single supporter of The Freedom Plan will fund enough money into the escrow account to build a new 1.5 MW $1.6 Million wind turbine every one to two years. That may not

---

[220] AWEA "Wind Power Outlook 2004" p. 2.

sound like much, but it is a terrific start, and one that the founders of that company can be proud of. The most difficult time for any new company is the beginning stage: once the foundation has been built, growing a company from an existing foundation of customers and sales reps is far easier than building a company from scratch. Krystal Planet is well on its way to becoming a billion dollar company – it is now only a matter of time. It is my hope that many other companies partner with Krystal Planet to endorse The Freedom Plan and accelerate its growth.

Each individual FutureWind℠ customer makes a big difference:

Pollution reduction. Every customer on FutureWind℠ for a year eliminates **172 grams of mercury** (poisons fish, linked to autism and other diseases, pollutes water) **67 pounds of nitrogen oxide** (smog, air pollution, asthma, headaches), **55 pounds of sulfur dioxide** (acid rain, eye damage), **12 tons of carbon dioxide** (global warming), **hundreds of pounds of solid particulate matter,** (air pollution, asthma, lung disease), plus hundreds of milligrams of lead, cadmium, arsenic, other heavy metals, and trace uranium, as well as dozens of pounds of poisonous carbon monoxide.

Build the Wind Turbine escrow account. Each month, at least $5-10 of the $30 FutureWind℠ Green Tag goes into the escrow account to build a new wind turbine (the range is determined by whether you waive the optional monthly thank you gift). For every total of $600,000 accumulated in escrow, a new wind turbine is built somewhere in your country, and likely in your region. Therefore, one person on FutureWind℠ for five years who waives the gift option contributes $600 to escrow. This represents one-tenth of one percent of the 1.5 MW wind turbine, which is enough to generate 1.5 kW per hour on relatively windy days, or up to 1,080 kWh per month for the 50-year life of the wind turbine. Since the average U.S. home consumes about 850 kWh every month, and your pro-rata share of the new wind turbine your five-year commitment helped build generates 1,080 kWh per month, you have effectively replaced your entire monthly consumption of electricity with clean, renewable, American-made wind energy – for 50 years!

Implement The Freedom Plan. Each new customer acquired by any company that has fully endorsed The Freedom Plan helps make the case to other customers that this movement is for real. And each customer is encouraged, and provided with a financial incentive, to help us find additional customers. Each new milestone achieved will be celebrated with great fervor, and the resulting media coverage will continue to add more credibility to the cause. Imagine the impact to readers of a newspaper when they read about a ribbon-cutting ceremony in their county or state of a new wind turbine, or wind farm, or sustainable building designed with wind and solar integrated into the structure. The popularity of reducing dependence on foreign oil and fossil fuels has never been higher, and should remain popular for many years. Then, imagine what their reaction might be when the reader of that article is asked to enroll on a Freedom Plan Green Tag a few days or weeks later by their neighbor, co-worker, friend or family member. They will certainly be more receptive to the notion of supporting the cause when they have already been reading about it in the paper or hearing about it on the news. Every wind turbine built as part of The Freedom Plan will make a difference – and will help us enroll even more customers in the area. And listing your name as a customer on websites or including you in milestone statements makes a significant impact on the collective credibility of the cause. Your commitment therefore adds significantly to the momentum already building to geometrically accelerate the implementation of The Freedom Plan.

Clearly, one person on FutureWind℠ or an equivalent future Green Tag **does** make a big difference.

As you will learn in the next chapter, The Freedom Plan affiliates intend to acquire one million futures Green Tag customers during Phase One of the business plan. During Phase Two, The Freedom Plan affiliates intend to grow their collective customer base 10 million residential subscribers and 1 million commercial Green Tag customers (with an average of roughly 3,000 kWh per month for each commercial account and millions of dollars transferred to the wind turbine construction escrow account every month).

The Freedom Plan calls for a total of 2 million megawatts of new wind energy generation. As described in Chapter 6, this 2,000,000 MW of wind energy will be installed initially at $1 million per MW (today's price) and drop rapidly in cost with the large volume of manufacturing production assured by the steady growth of those supporting The Freedom Plan. As we learned in Chapter 4, the cost of wind energy has dropped by 90 percent since 1980 with minimal growth in the wind industry. The U.S. has about 6,000 MW of installed wind energy and worldwide wind installations have barely passed the 50,000 MW mark. It should therefore be reasonable to assume that the costs will decline rapidly with a very large market of an additional two million megawatts. The costs to build new transmission, hydrogen production systems, utility-scale fuel cells and hydrogen storage are all included in The Freedom Plan with costs initially based on today's prices for all of these systems, declining appropriately with scale and mass production.

Affiliates of The Freedom Plan agree that the lion's share of the enormous revenues – and profits – generated by thousands of wind turbines producing electricity for hundreds of utility and industrial customers will be *reinvested into additional wind turbines* at new and existing wind farm sites throughout North America, as well as hydrogen production facilities, new and upgraded transmission lines, and hydrogen pipelines. I intend to make sure of that – as it is my mission in life to transform America's energy industry to clean power.

This self-fulfilling cycle of selling Green Tags to homes and businesses to build wind turbines and related clean power systems, using profits and revenues from those wind turbines to build more wind turbines while continuing to sell even more Green Tags, creates a potent formula for The Freedom Plan. And, although the math behind this formula is relatively simple, it leads to a commanding jump-start of The Freedom Plan: well over 25 percent of the total wind energy generation needed can be constructed by the primary affiliates alone within the first 10 years – even without any help from secondary affiliates that will certainly be supporting the Clean Power Revolution! Many other companies will undoubtedly contribute to the implementation of The Freedom Plan, and hopefully some of them will offer even more elegant techniques than those already discussed.

As the Clean Power Revolution's base of customers continues to grow, another benefit will be the ability to influence political parties by our collective voice and through mass communications through our grass roots network of like-minded individuals. From many voices comes one succinct message that can cut through the rhetoric we hear from politicians due to the well-funded energy industry lobby. This simple but compelling message will be broadcast to politicians across the land in written, verbal, and electronic communications: support The Freedom Plan – or be forever left behind. Although support from the government is not a prerequisite of a successful Freedom Plan, it can accelerate its completion.

In conclusion, by the year 2006 when The Freedom Plan affiliates hope to have one million residential customers and 100,000 commercial accounts, The Freedom Plan will be well underway:

- Based on $7.50 per month deposited into escrow for each residential customer and $20 per month for each commercial customer, $114 million will be added to escrow in a year. At $600,000 per wind turbine needed in cash equity (the balance being financed), that builds 190 1.5 MW wind turbines or 285 MW total.

- The following year, assuming modest monthly growth of just over 8 percent per month (much lower than the current 112 percent monthly growth rate) 570 MW of wind energy will be added.

- In 2010, by the time The Freedom Plan enters Phase Two, the assumptions can reasonably be modified slightly: each month, $10 goes into escrow per residential customer and $25 per commercial customer.

- During Phase Two (expected to be in five years by 2010), The Freedom Plan affiliates will have acquired more than 10 million residential customers and one million commercial customers. This assumes a growth rate that is much slower than that experienced to date, and one that I consider realistic and attainable.

- There are nearly 500 million people living in North America, so enrolling 64 million consumers and 10 million commercial entities on clean power in the form of Green Tags is not unreasonable, especially considering the impact it will have, and the consequences if we do not. Remember, the polls show that 75 million Americans are willing to pay more for clean power today, which does not include those living in all the other countries in North America. It also assumes that people are not offered a tax deduction, monthly gift, or referral program to make the Green Tags more attractive and expand the potential customer base. And, the number of people willing to purchase clean power will surely increase as energy costs continue to rise, the health and environmental damage of our energy policy become more well-known, and the security of our energy supply becomes more tenuous.

- With continued steady growth over just five years, *the The Freedom Plan affiliates could single-handedly implement The Freedom Plan simply by selling enough Green Tags*. The escrow account alone will construct over <u>75 percent</u> over the total wind turbines necessary for The Freedom Plan (1.5 million MW) assuming a steady decline in installed costs and an associated decline of the amount of cash needed in escrow for each lower priced wind turbine. Bear in mind that as The Freedom Plan affiliates' balance sheets, revenues and profits improve, more and more wind turbines can be constructed using balance sheet financing to cover the remaining <u>25 percent</u> of wind turbines and other systems needed.

The Freedom Plan will be implemented with or without help from the government and the energy industry. It will happen with or without me, as I am but one messenger, one cog on the wheel of this Clean Power Revolution that is far bigger than any one person or any one company. It is inevitable that this plan will move forward. It is not a matter of if; it is a matter of when. As you have seen in this chapter, every person who supports The Freedom Plan by purchasing Green Tags makes a significant difference.

INVESTORS WANTED

Ron Pernick, co-author of "Clean Tech: Profits and Potential" by Clean Edge, an Oakland, California market intelligence and publishing firm, says this: "For all the hype about the New Economy, a real, sustainable new economy is emerging around alternative energy technologies."[221] He goes on to say that "a small but growing number of forward-thinking investors recognize that several of today's emerging clean-tech companies will be the Microsoft's of the future." Today there is no market need greater than the demand for efficient, inexpensive, environmentally friendly energy.

Warren Buffet's Berkshire Hathaway, through MidAmerican Energy Company, is building a $323 million wind farm in Iowa. The Iowa Utilities Board has allowed a 12.2 percent Return on Equity for this project as part of the utility rate base. Private firms building wind projects obviously do not have their investment returns limited by state utility boards. Therefore privately funded wind energy projects can see returns on investment of 15-30 percent assuming 60 percent debt leverage or greater and efficient monetization of the tax credits and depreciation.

Stan Abrams of Nathaniel Energy Corporation says "all things considered, what if you could invest in an industry that plays as well on Main Street as it does on Wall Street, and brings Republicans and Democrats together? With clean energy, you can. And maybe you should."[222] There will be hundreds of companies building and joining the Clean Power Revolution. Many of them have not even been formed yet as of the time this book is being published – and many are not

---

[1] Energy Central News, "Clean Energy: Turning Potential into Profit" March 4, 2004.
[2] PennWell Publishing Company, February 2004.

yet publicly traded. Some of the privately owned companies may accept investors if they have filed appropriate offerings with the proper regulatory agencies such as the SEC (Securities and Exchange Commission). I recommend all of us who are supporting the Clean Power Revolution keep our eyes open for investment opportunities with young exciting companies in this sector. This is an area that will be seeking a great deal of new investment capital to accomplish the aggressive and necessary objectives of converting the world to clean power. Check with your investment advisor(s) and research any such opportunity thoroughly.

Your support of Green Tags and energy efficiency products and services, through Krystal Planet or another company that has publicly endorsed The Freedom Plan can only have one effect: it will accelerate the implementation of the greatest initiative this great nation has ever seen. One person clearly **can** make a difference. Not only does it feel good, *it is the right thing to do.*

Chapter 11:

AN EXCITING FUTURE

"I've come to believe that each of us has a personal calling that's as unique as a fingerprint - and that the best way to succeed is to discover what you love and then find a way to offer it to others in the form of service, working hard, and also allowing the energy of the universe to lead you."
Oprah Winfrey

The Moorehead municipal electric utility in Wisconsin discovered that the economics of wind power are now competitive enough to:

Fix power rates, and
Reduce fossil fuel generation.

Moorehead's green power program allowed customers to buy a subscription to the first planned wind turbine in advance of construction. Once sufficient subscriptions were in hand, construction could go forward using a form of customer-based financing similar to that proposed in The Freedom Plan. In the end, Moorehead sold *twice* the number of necessary subscriptions and consequently they built a second wind turbine. This is just one example of a market-driven technique creating a clean power project and which resulted in a greater economic justification for the project than originally expected.

The Freedom Plan is certain to be successful once enough people choose clean power in the form of Green Tags as described earlier. Many organizations from around the world are joining the Clean Power Revolution every month by endorsing The Freedom Plan and lending their support.

With such broad support, this plan cannot fail. The only two questions that still remain are: how long it will take to complete and who exactly will be participating? Obviously, one purpose of this book is to seek additional participants that will get involved directly in the Clean Power Revolution by generating excitement about The Freedom Plan. But this book is also intended to educate as many people as possible about clean power in general and the urgency of taking immediate action to save trillions of dollars. However, for those of you who:

Agree that choosing clean power is the right thing to do,
Want to know if doing so will really make a difference, and
Want to see The Freedom Plan successfully implemented, this chapter is for you.

We will show how steady growth of clean power customers guarantees success. And, we will see how long it will realistically take to accomplish our collective goal of absolute energy independence.

Phase 1

Objective: acquire **One Million** residential clean power customers.
Timeframe: 2-4 years.
What It Means: achieve **10 percent** of The Freedom Plan.

From a total of over 300 million Americans, and 75 million who the polls say are willing to choose clean power for a small premium each month, the objective is to obtain a mere one million customers in two to four years.

This phase is by far the most difficult, since getting this Freedom Plan train in motion from a dead stop is more difficult than accelerating a train that is already moving. Fortunately, Krystal Planet has already acquired thousands of customers in just a few months – so The Freedom Plan is already moving along at a modest clip. That makes it easier for other companies to join the crusade. An ancient proverb says, 'if you wish to eat an elephant, do so one bite at a time.' Therefore, we shall work backward from our objective of one million customers to show smaller step-by-step objectives. At each objective is shown the projected price of the Krystal Energy System (a turnkey home hydrogen system that can power the entire home and produce enough clean hydrogen to run two vehicles), as well as other notable milestones.

10,000 customers – construct the 1st two wind turbines; Krystal Energy System cost: $150,000.

25,000 customers – open the 1ˢᵗ Hydrogen fueling station, which provides clean, safe, renewably-produced hydrogen for automobile use, delivery to local homes for use in fuel cells as part of a *Hydrogen Community,* and for use in backup power applications; construct <u>five</u> more wind turbines; Krystal Energy System cost: $100,000.

50,000 customers - open <u>two</u> Hydrogen fueling stations, construct <u>10</u> more wind turbines; Krystal Energy System cost: $75,000.

100,000 customers - open <u>five</u> Hydrogen fueling stations; construct <u>25</u> more wind turbines; Krystal Energy System cost: $50,000; build the 1ˢᵗ prototype carbon-fiber Hydrogen-powered fuel cell car to be completely manufactured using 100 percent wind energy and powered with 100 percent renewable hydrogen.

250,000 customers - open <u>25</u> Hydrogen fueling stations; construct <u>100</u> more wind turbines; Krystal Energy System cost: $45,000; build <u>10</u> Krystal-powered fuel cell cars.

500,000 customers - open <u>100</u> Hydrogen fueling stations, construct <u>300</u> more wind turbines; Krystal Energy System cost: $40,000; build <u>100</u> Krystal-powered fuel cell cars.

1,000,000 customers - open <u>500</u> Hydrogen fueling stations, construct <u>1,000</u> more wind turbines; Krystal Energy System cost: $20,000; build <u>500</u> Krystal-powered fuel cell cars.

Along the way, Energy Consultants with Krystal Planet who reach the Vice President level and key performers from other companies will be eligible to open Hydrogen conversion shops to convert cars and trucks to run on hydrogen as well as sell Krystal-powered fuel cell cars to customers in the area. This will be similar to a franchise, where The Freedom Plan affiliates provide technical support, financing and training. The price of hydrogen powered vehicles is already affordable: an order of at least 100 Toyota Prius hybrid vehicles completely converted to run on hydrogen instead of gasoline would cost about $60,000 each in 2005.

Once a million customers have been acquired and 1,400 wind turbines (over 2,100 MW) have been installed by that time, The Freedom Plan affiliates will be able to build even more wind turbines every year. This will be accomplished through the use of a portion of profits from general operation, through leveraging the portfolio of existing wind projects already built by the The Freedom Plan affiliates which are generating revenue, through the affiliates' growing balance sheet, and through an estimated 100,000 commercial customer accounts. Combined, these factors will finance a total of <u>2,325 MW</u> of wind energy projects in the first year after acquiring the one million residential customers. This assumes average profit of only 20 percent of gross revenues, and assumes balance sheet financing of only 20 percent of the prior year's total installed MW of wind energy capacity. These are both conservative figures according to the industry standards. 2,325 MW is only 0.1 percent of the two million MW needed for The Freedom Plan, but it sets in motion the completion of 10 percent of The Freedom Plan.

Many exciting young companies in new fast-growing industries have grown at staggering rates over the course of history. Examples include Microsoft and Dell Computer in the computer industry, Cisco and Nortel in the telecommunications industry, and eBay and Amazon in online shopping. Assuming a modest growth rate of only 20 percent per year compounded (this is far lower than what is likely in this exciting Clean Power Revolution), the 2,325 MW will grow to over 200,000 MW (10 percent of The Freedom Plan total) during the 10th year. And, the final two million MW mark is achieved in the 17th year. Therefore, without any substantial assistance from government or outside parties, **simply growing the base of clean power customers can achieve 100 percent of the conversion of America to wind and hydrogen within 20 years**. The costs to build and upgrade transmission lines and add hydrogen infrastructure are included in this timeframe and are paid from profits of the wind energy projects. These profits from wind energy facilities are intentionally not included in the calculations used to finance more wind projects since they must be set aside for building the infrastructure necessary for large-scale integration of wind energy into the nation's energy system.

Keep in mind that The Freedom Plan is a grass roots movement, not a utility initiative or government program. In other words, its success is not bound by the number of customers a utility has in its territory. Nor is it bound by the available land for wind farms (as discussed in Chapter 4 there are hundreds and hundreds of thousands of acres of land already leased and permitted for construction of wind projects just waiting for investors to finance them). It is no bound by the whims of political spending decisions, access to capital for transmission lines, or availability of technology for electrolyzing water into hydrogen.

The Freedom Plan is self-funding and self-reliant. Its success is 100 percent dependent on market forces: when enough customers choose clean power through a company or non-profit that has endorsed The Freedom Plan (and is certified by an independent third party as a direct contributor to the wind energy escrow account as audited by a public accounting firm), the conversion begins to take place naturally. Conversion to 100 percent clean power is inevitable. There is no way The Freedom Plan can fail assuming enough homes and businesses choose clean power. Failure is not an option.

However, complete success in 17 years is not fast enough. The objective is a 10-year completion in order to save America the additional trillions of dollars of costs outlined in Chapter 7 caused by the extra seven years above our 10-year goal. Therefore, Phase Two adds a catalyst to the process once Phase One has achieved its primary objective of one million residential customers.

Phase 2

Objective: **10 Million** clean power customers + **1 Million** H2 vehicles + **1 Million** commercial clean power customers.
Timeframe: 4-6 years after completion of Phase 1.
What It Means: achieve **90 percent** of The Freedom Plan.

Phase One was accomplished with very little outside capital investment, with the exception of using conventional bank and equity financing of each wind energy project. But the calculations assumed no large infusion of public investment in the companies building the core components of The Freedom Plan. Its success in 17 years was based solely on financing the build-out from revenues and profits.

A public offering of stock in some of The Freedom Plan affiliates may or may not ever take place. But if a public offering or an attractive private equity investment were to ever take place, it could potentially raise enough capital to substantially increase the number of MW installed each year. In fact, a successful public offering during the latter years of Phase One could raise enough capital to shave two or three years off the 17 year completion timeline as shown for Phase One. This suggests a completion date in the 14th or 15th year after commencement (the year 2019 or 2020, since this movement has already begun and the first full year is 2005). It also shortens the goal of achieving 10 percent of Freedom Plan by three years to the 6th year (2011). *Disclaimer: there is no promise or guarantee in this or any other statements made in this book that Krystal Energy, any of its subsidiaries, or any other company supporting The Freedom Plan will ever go public.*

If such a public offering were approved by the Krystal Energy Board at some point during the final years of Phase One (or by any other company endorsing The Freedom Plan), it would save trillions of dollars of the costs of our current energy policy as described in Chapter 7, including the extremely high cost of extracting the last couple decades' worth of oil from the dwindling reserves. Krystal Energy may or may not go public, as it will not make a significant impact on shortening The Freedom Plan timeline. However, there appears to be a very strong demand by the investing public to own a piece of socially responsible clean power companies, especially in the fastest-growing wind sector where few public companies exist today. Hence, at the appropriate time the Krystal Energy Board, as well as the Boards of other related companies supporting The Freedom Plan, may consider such an option.

However, one of the best ways to truly accelerate The Freedom Plan implementation is to tap the automobile market. By focusing on the production of renewable hydrogen for the transportation industry, a *new market* for wind energy is created. This market is enormous and lucrative, assuming enough grass roots entrepreneurs are available to help convert existing vehicles to run on hydrogen. There are already countless companies developing fuel cells for vehicles, and many already have working units that are nearing commercialization. Assuming a widespread market for hydrogen was available, these companies could rapidly scale up manufacturing to produce low cost, extremely efficient hydrogen powered cars and trucks whose customers would buy large quantities of renewable hydrogen.

The Phase Two objective of One Million hydrogen powered vehicles will spur additional construction of enough wind energy facilities and related hydrogen production infrastructure to support this demand. The growth from this one new market will start slowly in year three (2007) and then add hundreds of thousands of MW of new wind energy installations by year ten (2015) in a rapidly scaled up timeframe (see Timeline chart in Chapter 6). This achieves the objective of full-scale construction of the entire Freedom Plan within 10 years, by 2015, solely using market forces based on customer demand for clean power.

Phase 3

Objective: **30 Million** residential clean power customers + **10 Million** H2 vehicles + **3 Million** commercial clean power customers.

Timeframe: 1-2 years after completion of Phase 2.

What It Means: achieve **100 percent** of The Freedom Plan.

At this point, The Freedom Plan is fully implemented. Krystal Energy Corporation and the other successful companies who helped achieve this great victory will be the darlings of Wall Street if publicly traded. They will be, whether public or private, some of the largest companies on the planet – if not the largest. They will be the most respected companies on the planet. And they will be sustained by a network of hundreds of thousands of entrepreneurs, Energy Consultants (in the case of Krystal Planet), volunteers, non-profit organizations who have been actively supporting the Clean Power Revolution, and other key stakeholders. The ownership of these public and privately held companies will likely be widespread, since nearly all of them will likely offer registered stock option programs to their employees and/or key contractors. Therefore, the profit and wealth created by this massive movement will be more widespread than in any other movement in history. This prevents the control of the Renewable Age from falling into the hands of any particular small group of wealthy, influential politicians or powerful companies. And this is as it should be. This shift should be created by the people using grass roots marketing. And those who create and support this vital shift of wealth in the largest industry in the world should be the recipients of the profits and wealth created. This single factor alone could create more millionaires and billionaires than any other single movement in history. The completion of The Freedom Plan will save our great nation trillions of dollars of unnecessary spending related to fossil fuels. And it will ultimately lead to much lower overall energy prices since the fuel cost to operate these renewable energy facilities is essentially free.

Krystal Energy Corporation and its affiliates, which again will most likely be owned by millions of people across the planet, will become a fully integrated energy company. It will own wind farms, solar farms, hydrogen fueling stations and pipelines, hydrogen electrolyzers, desalination plants (to make fresh water for use in city drinking supplies and for use in the electrolyzers near oceans), transmission lines and other supporting infrastructure and software systems. It will own the generation assets to produce clean power, the transmission and distribution assets to move the power and hydrogen around the world, and the retail clout to sell hydrogen conversion kits for cars, Green Tags, fuel cells, vehicles, and other clean power products. This will undoubtedly mean that Krystal Energy Corporation will become the largest owner/operator of renewable energy facilities in the world. And it will likely surpass the $100 Billion per year mark shortly before completion of The Freedom Plan and perhaps become the first in the world to become a Trillion Dollar company.

During Phase Three (or perhaps in Phase 2), it makes sense for Krystal Energy to purchase a leading fuel cell manufacturing company to become more vertically integrated. Fuel cell manufacturing costs, like other manufactured products, should drop by 15-30 percent for each doubling of cumulative production units until limited by the cost of basic materials.[223] For example, if a clean power company were to produce 100 fuel cell cars in 2006 at a cost of $100,000 each, the next 100 cars should cost no more than $85,000, and by the time 1,000 cars have been produced the average cost should be well below today's price for a premium vehicle – and could easily last up to 20 years if powered by clean hydrogen and build using lightweight rust-proof carbon fiber.

It may also be prudent to launch an automobile manufacturing company (or purchase a forward thinking, open-minded existing automobile company) to manufacture hydrogen-powered carbon fiber vehicles within one company. There may even be unique technologies that such an automobile company would be willing to embrace due to the sustainable philosophy found among the family of Freedom Plan companies. An example of one such technology would be to capture carbon molecules out of carbon dioxide found in the air (the leading global warming gas), leaving only the oxygen to escape back into the atmosphere where it is needed. This technique could collect all the carbon necessary to manufacture all the carbon fiber chassis (automobile frames) and auto parts needed for large scale manufacturing of new hydrogen powered vehicles. Carbon fiber used in place of heavier and more energy intensive steel as a structural material that is derived from the air would make cars that are lighter and far stronger than steel frames, but would also remove carbon dioxide from the air for every car sold!

Another likely result of reaching Phase Three will be the acquisition (friendly or hostile takeover, depending on the case) of old-school, slow-to-change electric utility companies. A reason for doing so would be to buy their assets (including transmission lines) and shut down their coal, gas, and nuclear plants while replacing those 19th and 20th century power plants with 21st century clean, safe, renewable power generation from wind, biomass, solar, geothermal, renewable methane and hydrogen. Many of these electric utilities are publicly traded making them easier to acquire. And, many of them are struggling to remain profitable (some are even seeking bankruptcy protection due to high fossil fuel prices and the high cost of meeting new emissions standards being required of them) so they are actually seeking new capital and investors. In theory, The Freedom Plan could be implemented quite successfully even if utility companies are not cooperative. However, in reality most utility companies and other

---

[223] H. Tsuchiya & O. Kobayashi ("Fuel Cell Cost Study by Learning Curve," EMF/HASA International Energy workshop, Stanford University, 18–20 June 2002) predict range of 14–26% per doubling.

energy companies would be eager to purchase clean power from an independent power producer in their region offering energy at less than three cents per kilowatt hour. Therefore, the successful completion of all The Freedom Plan steps outlined in Chapter 6 is virtually guaranteed with enough customer participation.

There is always hope. And now that a plan is in place to convert our nation to clean power and save trillions, all we have to do is go sell the idea to enough people and get them to choose clean power.

CLOSING THOUGHTS ABOUT OUR FUTURE

Your home can become its own clean power plant. Phase One is about enrolling people on clean power while selling energy saving products and services to cut energy demand by at least 20 percent across the board. It includes some hydrogen communities in selected regions where enough customers have enrolled, but Phase Two actually includes widespread sales of home hydrogen systems. This will essentially take thousands of homes 'off-grid' across the nation, cutting electric demand significantly as a part of this distributed generation component of The Freedom Plan. These products will be sold by various entrepreneurial companies but will not be paid for as a part of The Freedom Plan. That is because homeowners will want to spend the money to have the freedom of never paying a power bill again and finance it themselves. This further adds to demand for hydrogen, electrolyzers, and fuel cells, which will accelerate the drop in costs to manufacture all these products, add more competition to the marketplace (further dropping costs and shortening product development times), and dramatically accelerating the need for more renewably produced hydrogen. This factor was only slightly included in the timeline schedule to remain conservative, but could actually shave another 1-2 years off the total time it takes to complete The Freedom Plan.

I am a proponent of both energy independence and financial independence. That is why I left the conventional wind development business to form Krystal Planet: to allow anyone in the world to save money on energy using unique products, and earn enough money by referring others to the Clean Power Revolution to have 100 percent of their energy costs paid for through savings and/or referral bonuses. As new products become available, we intend to have one of the largest distribution networks on the planet to promote such products to the world and further lower consumption of energy.

For example, one of the most common energy products is one of our most inefficient. Using incandescent light bulbs, which were invented in the 1800s and have not changed much in over 100 years, is foolish. Compact fluorescent light bulbs, which use 80 percent less power and last up to 15 times longer (you rarely have to replace them) are one product that everyone should buy regularly. But many people just do not know they exist or why it makes sense to pay more for a light bulb (someone needs to educate the consumer that they will save far more money in energy costs and buy far fewer replacement bulbs). Other items will also need some explanation, like an under counter probiotic digester for food waste. This product can prevent millions of tons of organic waste going into landfills, creates a terrific potting soil for your indoor or outdoor plants, and could eventually create renewable methane in your own home to blend with your natural gas and/or hydrogen for heating. Products such as these can – and will – be promoted as part of the Clean Power Revolution by those companies who officially endorse The Freedom Plan.

Krystal Planet, as you have learned, is not just a vehicle for energy savings and choosing clean power. It is also a vehicle for anyone in the world to achieve financial independence by making the Clean Power Revolution a career. There will be many other companies joining Krystal Planet in support of The Freedom Plan, and some of those may also have lucrative options for individuals or small business owners to earn income by selling clean power products and services. I welcome them all! We need all the support we can get to make The Freedom Plan a reality. We are all on the same team here, because America must reduce her dependence on fossil fuel and foreign oil. And then, we should lead the world towards a further conversion of the entire planet to clean power, since this is a global issue that affects us all. What happens in China and Europe and South America makes an impact on our oceans, air, and food supply worldwide.

The good news is that we now have a solution before us that is realistic and attainable, and is guaranteed to work if we just get enough of our friends, neighbors, and citizens to choose clean power through a certified Freedom Plan organization. I encourage you to do your part – do the right thing – and find a certified organization you like, and choose clean power from them today. We need your help to complete The Freedom Plan – and help **you** save thousands of dollars over the coming years. There is only one reason that The Freedom Plan might fail. That is this: if you do nothing, if you do not actively seek a clean power choice, The Freedom Plan may fail. That would be a very, very costly failure that none of us want to live with.

Chapter 12:

CONCLUSION

If a single solution exists to wean our nation from the stranglehold that oil and fossil fuels have around our collective economic neck, then that solution should be emphatically endorsed by the masses. One of the primary challenges of creating significant change in the energy industry is the fragmentation of various interests.

For example, there are numerous environmental groups who have been working for years toward the very respectable goal of cleaning the air. Unfortunately, their impact is diluted since there are so many different groups working towards the same general goal. These groups include non-profit organizations, countless consumer associations, public interest groups, quasi-governmental agencies, and many more. The overall success they have enjoyed, while impressive in many instances, is substantially reduced due to fragmentation. These multiple groups have varying clean air agendas and success targets, different priorities based on sources of funding, and poor communication with each other. Once the Clean Cities program was initiated by the U.S. Department of Energy, the overall national results began to improve significantly. This single national program helped form local networks of all the groups working on clean power to facilitate better communication, create consistent national goals, and identify pressing local objectives that each local group could rally behind. Although there is still a great deal of wasted effort due to duplication of projects and resources, the overall effect of national goals and a national program has been quite positive.

Analogous to the example above, there are many organizations working toward clean power. Some focus on renewable energy such as wind or solar power or both. Some focus on clean air and water. Others spend their time working towards reducing dependence on foreign oil (ethanol and bio-diesel organizations are good examples). Many have other primary objectives, but support the idea of clean power philosophically. Some of these organizations are non-profits, while others are businesses hoping to make some money doing a good thing as a part of the anticipated worldwide conversion to clean power. And finally, many citizens and business would wholeheartedly embrace the idea of clean power if they knew their support would make a difference and not just be considered an ambiguous 'donation' to a good cause.

Furthermore, wind development companies compete fiercely with each other for the 'Holy Grail' of the wind business: the power purchase agreement (PPA) from a creditworthy electric utility. These contracts are extremely difficult to obtain, are doled our sparingly by utilities, and make financing new wind projects a time-consuming, very expensive proposition. Therefore, the wind companies that are all trying to compete for these coveted agreements pull out all the stops to attempt to obtain them and rarely – if ever – cooperate with one another to simply put more wind turbines in the ground. Financing a wind project almost always requires a PPA. But if these wind companies knew there was going to be an enormous influx of capital through a massive, well-orchestrated clean power movement, they would instantly become more cooperative with each other. They would understand that there would be plenty of capital to go around to finance thousands of wind projects, so all wind developers who locate and prepare adequate sites would get a piece of the pie. There would be plenty to go around for everyone.

The same logic holds true for other renewable energy technologies. Wind energy competes for limited clean power dollars with the other renewables such as solar, geothermal, biomass, landfill gas. With a clean power revolution organized around a single common objective, all these companies would get a piece of the action. This provides a nice incentive for each of these companies to work with one another to develop potential projects which could encompass multiple technologies, such as wind and solar working together at the same site to increase the overall output and efficiency of the project, or biomass and wind at an ethanol plant, or landfill gas and solar power at a city dump.

Therefore, these scattered, fragmented companies, organizations, and groups across the country are quite ineffective relative to the potential. If all these parties were to unite around one common objective, imagine the economies of scale that could be gained. One clearly defined cause that included something for all these parties, wrapped up in a well-organized effort, could accomplish far far more than the sum of all of the individual parts.

That cause, of course, is The Freedom Plan. Instant access to key information in real time, rapid global communication of all aspects of the plan including progress updates and new technologies, a database including all participants to facilitate cooperation and information sharing, and massive purchasing power as a result of the sheer size of the project are readily provided by The Freedom Plan. I know this is my calling, and many others with whom I have the privilege of working also feel this movement is their life purpose. And there have been so many surprising positive events that occurred at just the right time, or the perfect person contacted us just when needed to assist our efforts along the way it is my personal opinion that these inexplicable events are actually a sign that we are on the right path. It has been an honor to be a part of this incredible cause thus far, and I look forward to spending the rest of my life converting the world to clean power. The many new friends I have made as a part of this Clean Power Revolution are also committed to this movement. It is sure to be a fun ride over the next ten years, working with people, companies, and organizations to save the planet while making money and having a good time.

It will be gratifying watching participating non-profits earn revenue by helping to promote clean power, using The Freedom Plan as a fundraiser as well as a great 'feel good' supplement to their primary objective. Companies participating in The Freedom Plan can also become good corporate citizens by supporting such a terrific cause. They can also help build shareholder value and mitigate risk by supporting a) the diversification of the energy supply in their region, b) the injection of clean, renewable power into the system which should in turn lead to lower energy costs, and c) the elimination of higher future operating costs (in the form of higher taxes, health insurance premiums, and substantially increased costs for supplies and raw materials). This increased cost of doing business is inevitable as described in Chapter 7 if the nation does not quickly convert to cleaner sources of power.

Can it be done? Randall S. Swisher, Executive Director of the American Wind Energy Association, says in a report called "Bringing Wind Energy up to 'Code' " that in Europe, where concentrations of wind energy are the highest in the world, wind turbines now operate with a variety of features that actually enhance grid reliability. The U.S. Deputy Secretary of Energy Kyle McSlarrow says the goal of 100,000 MW of wind power by 2020 (6 percent of U.S. electricity needs) is realistic.[24] This is coming from a government representative of a conservative Bush Administration. That should tell us that with an exciting new plan that relies on market driven forces that a grass roots campaign can initiate without relying on government support, we can do far better.

---

[24] Presentation at Global WINDPOWER 2004, Chicago, March 29, 2004.

The United States currently spends 11.2 percent of its GNP (gross national product) on energy; Japan spends only 5 percent. This is due to inefficiency, according to Arthur H. Rosenfield, director of Center for Building Science at Lawrence Berkeley Laboratory. He estimates that this costs the U.S. $220 billion per year and gives a 5 percent advantage to Japan on everything they sell.

There are 121 countries that have endorsed the Kyoto treaty on global warming. With the latest addition, Russia, the treaty will now formally go into effect. The United States is responsible for 36.1 percent of worldwide emissions of greenhouse gases,[225] and yet the U.S. has not endorsed the Kyoto treaty. As discussed in this book, the economic windfall (pun intended) of endorsing this treaty is stunning. Any politician or economist who tries to convince you that the economic cost of endorsing the pact would damage the global economy is simply not aware of (or willing to acknowledge) the facts. The cost to our nation of *not* endorsing it and converting to clean power will cripple our country's economy and could, over several decades, ultimately cost us our lofty position as the world's lone superpower. The Soviet Union went broke by attempting to keep pace with U.S. military technology spending using an inefficient economic system and archaic political management. They could not sustain the huge trade and government spending deficits.

Uncle Sam currently suffers from the largest government spending deficit in the history of our nation with no end in site. And, the U.S. is losing hundreds of billions of dollars per year as the result of the worst trade deficits in our nation's history. Combined, these two deficits are causing the dollar to weaken substantially relative to the Euro. Europe is a region of the world far ahead of the U.S. in reducing dependence on foreign sources of energy and fossil fuels which cost billions in health and environmental damage. If the United States clings to an archaic and costly energy policy and continues to spend heavily on military action and technology to 'protect national security interests' in oil-rich regions of the world, it is certainly possible for this inefficient use of resources in our largest industry to eventually bankrupt this great nation.

The Europeans are kicking our tail in building clean power projects, especially wind and hydrogen. Will they lead the world in the Clean Power Revolution, and leave us Americans in the dust? This wakeup call to the world presents an exciting opportunity to America. We are losing this call to lead the world – and losing badly. But we have the greatest renewable energy resources in the world, and the greatest entrepreneurial spirit. I believe we are up to the challenge. We

---

[225] TIME Magazine, October 11, 2004, p. 25.

can – and should – capitalize on this once in a lifetime opportunity. Those of you readers who have already made the decision to stake your claim on this massive shift of wealth to clean power should be encouraged. An exciting, attainable plan exists to follow, and it includes a comprehensive support system. All you have to do is follow it with passion and energy and never quit.

According to a May, 1998 Fortune magazine article, "Only a third of U.S. manufacturers are seriously scrutinizing energy usage." That was right about the time that energy prices began creeping higher. Analyst David Blackburn, Professor Emeritus of Economics at Duke University, **believes inefficiency accounts for 50 percent of the energy used in the U.S.**[226] Therefore the principle of energy efficiency, an integral component of The Freedom Plan, could potentially cut America's total energy consumption of 10,000 TWh per year to 5,000 TWh. The Freedom Plan calls for building enough renewable energy production to provide for 8,000 TWh per year, or 80 percent of the total. This allows for only modest success in the energy efficiency portion of the plan to remain conservative throughout.

Bill Gates, the founder of Microsoft, saw an opportunity. When he did, the secret to his early success in a new fast-growing industry was simple: he took immediate and massive action. Now is the time for you to learn from Bill. Take immediate and massive action to launch your own clean power business, whatever that may be. Over the next ten years, the entrepreneurs (including Krystal Planet Energy Consultants) who join the Clean Power Revolution and build their business with fervor will create an absolute dynasty for themselves and their families. Many of these individuals will become millionaires and create so much wealth their family will be financially independent for many generations. A few hardworking Ecos and other business owners contributing to the milestones of The Freedom Plan will become fabulously wealthy, actually becoming billionaires. In fact, there will be hundreds of new billionaires and tens of thousands of new millionaires created in the Clean Power Revolution.

In the only industry on the planet that measures its trends in trillions as opposed to billions, this enormous industry is ripe for new blood and new visionaries. The leaders of this exciting industry will create fortunes with the many companies implementing The Freedom Plan. It is my personal goal, as one of the spokespeople for the Clean Power Revolution and the leader of one of its leading companies, to create at least 10,000 new millionaire families and 10 new millionaires through Krystal Planet. That sounds unbelievable; even as I write the

---

[226] http://www.mecgrassroots.org/NEWSL/ISS38/38.07CostNuclear.html.

words I can hardly believe what I am saying. But the sheer size of the industry and the magnitude of the opportunity provide unprecedented wealth-building potential. It is mathematically impossible for this shift to clean power to occur without creating enormous new wealth for those pioneers who lead the way.

You have a choice in front of you now. It is a simple choice, really. You can continue supporting dirty power. You will be contributing to poor health and asking for an average of $750 per month ($9,000 per year) of increased costs for the next 20 years. Or, you can do the right thing and choose clean power. You will save yourself a bundle of money over the next two decades, clean up the air you breathe, the water you drink, the food you eat and the increase the health of your body and those of your family. You can even make a few bucks along the way if you choose to, simply be joining this 10-year 'Odyssey' that has already been set in motion. This may even be your destiny to join this cause and become a leader of the Clean Power Revolution in your area.

It really is that simple. How bright our future will be depends on one simple choice. I sincerely hope you choose to do your part and support The Freedom Plan. Years from now, when your children and grandchildren ask whether you supported clean power when you had the chance, you will be glad you did.

Making the right choice will not only save the planet, it will save **you** and your family hundreds of dollars per month and thousands of dollars per year – for the rest of your life. Choose clean power today and go forth in peace, knowing forever that you did the right thing.

ABOUT THE AUTHOR

Written by wind energy expert, popular public speaker, and published author Troy Helming: www.TroyHelming.com

Troy Helming was born in 1967 in Denver, Colorado. He and his wife Alysia live in Lenexa, Kansas, in a home that is 100 percent pollution-free. Troy and his wife drive cars that are 100 percent emission-free.

Troy Helming is the Founder & CEO of Krystal Planet Corporation, which is a grass-roots effort to enroll 1 Million homes on clean power by Earth Day 2007. Krystal Planet is located at 8527 Bluejacket Street, Lenexa, Kansas 66214. Its offices are 100 percent pollution-free, powered by clean, renewable, American-made wind power.

Mr. Helming has published numerous wind energy articles including:

1. "How to Finance a Wind Farm" Bank News Magazine, March 2003
2. "Earth Day & Iraq: a Connection?" published in various newspapers
3. "Uncle Sam's New Year's Resolution" as published on www.RenewableAccess.com and The Sustainable Investor

Mr. Helming has been featured in each of the following media:

PBS
NPR (National Public Radio)
Fox News 4
The Discovery Channel
Ecotalk Radio on Air America
Kansas City Small Business Monthly
Talk Radio 610 a.m.
The Kansas City Star
Sky Radio (on airline shows aired on flights with American Airlines, United Airlines,
    Northwest, Delta, and others)
Forbes Magazine
www.Forbes.com
Network Marketing Business Journal

As a public speaker and wind energy expert witness, Mr. Helming has been asked to present to audiences in dozens of states in the USA from Hawaii to New York, and Florida to California. His travels have also taken him to numerous countries to speak about wind energy and The Freedom Plan.

Mr. Helming is a member of Mensa, Sigma Phi Epsilon, and the University of Kansas Alumni Association. He is an avid skier, practices Krav Maga martial arts and yoga, exercises regularly (running, swimming, weight-lifting) as part of a strict Wellness Lifestyle, and is working towards climbing the Seven Summits (the world's highest peaks on each of the seven continents).

To request Troy Helming as a speaker at an upcoming event, go to www.troyhelming.com or call 913-888-0500 x157. Please provide at least 60 days advance notice. Travel expenses are to be paid by the event organizer, and a speaker's fee may apply.

Please send your suggestions to improve this book in subsequent printings to troy@troyhelming.com or fax to 913.888.0551.

Certified Supporters of THE FREEDOM PLAN

Krystal Planet Corporation
Krystal Energy Corporation
Pristine Power Corporation
Save the Planet USA, a 501(c)(3) non-profit
Team Activate, LLC
SustainableMarketing.com

Last updated April 1, 2005

For the most current list, go to www.TheFreedomPlan.org.

To be listed as a certified supporter of The Freedom Plan, simply submit a letter of support to the author or the editor (see below). A formal review of your organization's support of The Freedom Plan will be conducted by an independent 3rd party to certify your organization as an official supporter of The Freedom Plan. There is no charge for the review, which may take up to 60 days.

Author: Troy A Helming troy@troyhelming.com.

By mail:
The Freedom Plan
c/o Troy Helming
8527 Bluejacket Street
Lenexa, KS 66214

143421